THE LONG SPACE AGE

THE LONG SPACE AGE

The Economic
Origins of Space
Exploration from
Colonial America
to the Cold War

Alexander MacDonald

Yale UNIVERSITY PRESS
New Haven & London

Published with assistance from the foundation established in memory of James Wesley Cooper of the Class of 1865, Yale College.

Yale University Press books may be purchased in quantity for educational, business, or promotional use. For information, please e-mail sales.press@yale.edu (U.S. office) or sales@yaleup.co.uk (U.K. office).

Set in Galliard Old Style and Copperplate 33 BC types by Westchester Publishing Services.

Printed in the United States of America.

Library of Congress Control Number: 2016952701
ISBN 978-0-300-21932-6 (cloth : alk. paper)

A catalogue record for this book is available from the British Library.

This paper meets the requirements of ANSI/NISO Z39.48-1992 (Permanence of Paper).

10 9 8 7 6 5 4 3 2 1

FOR THOMAS AND SUSAN MACDONALD

CONTENTS

ACKNOWLEDGMENTS

Although short compared to the journeys undertaken by some of the explorers of the heavens covered in this book, the path to the finished product here has included its fair share of wandering and exploration. The work for this book—begun in Canada, furthered in the United Kingdom, and finished in the United States—has been conducted at more libraries, coffee shops, cabins, hotel rooms, and apartments than I can remember. I have been fortunate, however, to have been propelled along the way by the support of a large number of mentors, friends, and colleagues, and I would like to hereby acknowledge their indispensable assistance.

I first and foremost acknowledge the support of General Pete Worden, who shaped and encouraged my nascent interest in this subject from our first meeting in Vancouver, Canada, over a decade ago and later allowed me an opportunity to experience the complexities of American space exploration firsthand. I cannot thank him enough for his support, kindness, and friendship. I am also forever indebted to Professor Avner Offer for his sage advice over many years, which helped structure my investigation and argument. Whether because of or in spite of his view that space exploration is an example of resource misallocation from a public-choice perspective, his advice was particularly insightful, and I am thankful to have been able to benefit from his wisdom and wit.

Discussions with Professors Knick Harley, Roy MacLeod, Charles Fox, Robert Grant, Jane Humphries, Allan Chapman, David Edgerton, Nicholas Dimsdale, and Keith Mason helped set the mold into which this effort was ultimately poured. For Alan Green a special thanks is reserved, for it was his lectures, insight, and dedicated mentorship that initially drew me to the field of economic history. He was also the first—and very nearly the only—economic historian who supported the idea of investigating the economic history of the space program. Thanks are due to Frank Lewis, Marvin McGuinness, Ian Keay at Queen's University in Ontario, and Mauricio Drelichman at the University of British Columbia, who all helped guide my early interest in economic history. The support and encouragement provided by Tom Reilly, Jackie and Mike Bezos, and the TED Fellows program have also been truly incomparable.

I would especially like to acknowledge the many friends and colleagues that have sharpened thought and provided new avenues for research over the years through our conversations. Jacob Foster, Sean Gourley, John Karcz, Pete Klupar, Creon Levit, and Kevin Parkin all stand out as having provided particularly valuable insights and critique. Special thanks go to Holly Loubert for providing helpful feedback and early inspiration. I am also especially grateful for the companionship, encouragement, and enthusiastic editorial pen of Morgan Matson.

The quick wit and friendship of Susan MacTavish Best has also provided much support, including the use of her fine cabin in Mill Valley, which was the scene of a critical writing phase. For Eric Berlow I am also thankful, as he kindly allowed me the use of his peaceful retreats, both in Wawona and Swall Meadows, where the book picked up its final momentum.

Another major debt of gratitude is also owed to the many librarians and scholars who have helped make this work possible. Librarians who have helped with this research include Fordyce Williams of the Clark University Archives and Special Collections, the staff of the Caltech Archives, the staff at the NASA Headquarters Archives, and the staff at the NASA Ames Research Center Library. Advice and encouragement from a number of NASA historians has also been helpful, most especially that of Bill Barry, Glenn Bugos, Steven Dick, Christian Geltzer, and Roger Launius. I am also immensely grateful for and humbled by the foundation of scholarship produced by the many scholars of economic history and the histories of astronomy, technology, and spaceflight on which this work has been built.

Wholehearted thanks go to Joe Calamia for his enthusiasm and encouragement of the project as editor, to Kate Davis for her indefatigable copyediting and fact-checking, as well as to the rest of the staff of Yale University Press who helped transform the initial manuscript into the final product.

Finally, I am forever indebted to my parents, Tom MacDonald and Susan MacDonald, for having taught me so much—including a love of history and exploration—and for having provided such unsurpassed support for so many years.

THE LONG SPACE AGE

INTRODUCTION

And what would be the purpose of all this? For those who have never known the relentless urge to explore and discover, there is no answer. For those who have felt this urge, the answer is self-evident.

—*Hermann Oberth,* Man into Space, *1957*

The rise of private-sector spaceflight and American billionaires pursuing their ambitions in space seems to be a new phenomenon. After the origin of space exploration as a government enterprise during the Cold War Space Age, entrepreneurs and individuals have become a new force on the scene and are increasingly the drivers behind some of the most prominent space activities. In the Cold War, the United States and the Soviet Union developed intercontinental ballistic missiles to deliver their nuclear warheads, creating the technology for satellites and spaceflight vehicles. The race into space then became an important dimension of the Cold War as the two superpowers competed vigorously to be the first to claim prestigious spaceflight achievements, culminating in an American victory with the successful expedition of Neil Armstrong and Buzz Aldrin to the surface of the Moon. After the unmatched success of the Apollo program, with no political need for further spectaculars, NASA was downsized, spaceflight was confined to low-Earth orbit, and further exploration was confined to robots. Since then, NASA spaceflight projects have continued to advance our knowledge of the solar system and the universe but have also become associated with cost overruns, expensive failures, and the deaths of fourteen American astronauts in the space shuttle *Challenger* and space shuttle *Columbia* accidents. With the Cold War over, the space shuttle retired, and no political urgency for major new space exploration activities, the momentum of the Space Age appears to have run its course, and we are entering a new era in which space exploration may become for the first time the domain of private individuals.

This is the conventional narrative of American space history in the media, as well as in historical scholarship.[1] It is also a narrative that results from an almost exclusive focus on governmental activities and thus misleadingly equates their relative rise and fall over time with the long-run history of American space exploration as a whole. There are indications that space exploration may be undergoing a period of revitalization, driven in part by the investments and motivations of wealthy Americans and entrepreneurs and in part by institutional transformations within government space programs. Investments by prominent contemporary American billionaires, such as Elon Musk and Jeff Bezos—each of whom has established his own privately held spaceflight company—have created new space capabilities, often in coordination with governmental programs and ambitions but also driven by their personal capital and motivations. Although these efforts are too recent to be subjected to historical analysis, it was from an interest in understanding the historical origins of private-sector space exploration that this book originated, and it is in hopes of better understanding the potential and pitfalls of those activities that it aims to provide a long-run historical perspective. These private-sector initiatives have been deeply contentious within the American space community, and the argument continues over their relative importance and appropriate role in the formation of national space policy. Regardless of one's personal views on the desirability of private-sector spaceflight capabilities, if we want to understand their emergence within their broader historical context, we need to examine the long-run economic history and relative importance of private funding in the development of American space exploration.

Again, according to the conventional narrative, it was public investment that created the Space Age, with private companies serving principally as government contractors and with private investment largely eschewing space exploration activities in favor of communications and remote sensing satellite applications. Although several billionaires, entrepreneurs, and private equity investors have, since the 1980s, devoted their intellectual efforts and money to the development of private spaceflight capabilities, their investments pale in comparison to NASA's billions of dollars of public expenditures on similar endeavors over the same period of time. The overwhelmingly governmental history of the Space Age within the context of the conventional narrative thus seems to confirm that private-sector space efforts are a novel and marginal phenomenon in the American space enterprise.

In this book, I will argue that this typical narrative is misleading; if we look at the history of American space exploration on a longer timescale, a very different history emerges—one in which personal initiative and private funding is the dominant trend and government funding is a recent one. The long-run history thus turns the conventional wisdom on its head: it is the governmental leadership of space exploration that is the more recent phenomenon, while the resurgence of private-sector space efforts in the early twenty-first century represents a return to an earlier pattern. This is the perspective that emerges when we frame the conventional mid-twentieth-century Space Age as only one phase of a Long Space Age that stretches back to the discovery of the telescope.

From one perspective, this book continues work by people such as science writer Willy Ley, who, in his seminal 1950 *The Conquest of Space,* made astronomy and spaceflight part of a unified narrative, with the spaceship presented as the next logical step in the progress of astronomy.[2] But within the context of the analytical literature on space history and space policy, this book is unique in both its long-run view and economic perspective. I believe this perspective is justified: the American astronomical observatories of the nineteenth century are considered as projects from an earlier phase of American space exploration, effectively equivalent in motivation and purpose—and in relative economic importance within their respective historical contexts—to robotic space missions to the planets in the twentieth and twenty-first centuries. Although the technology involved is different, they are each relatively complex, capital-intensive projects motivated by desires to explore the heavens. The telescope and the spaceship are thus, in this view, both instruments of space exploration. Examining these endeavors together as part of a continuum of space exploration activities—and using a consistent set of metrics to compare the costs of these space exploration instruments and projects—allows us to identify the underlying economic patterns and trends that remain obscure in space histories that are confined only to spaceflight during the unique circumstances of the Cold War and its aftermath.

A comprehensive history of the Long Space Age would be the work of a lifetime, so I emphasize three pivotal moments and movements in the development of American space exploration that are spread across two centuries. I chose the subjects to be specific enough to allow for contributions to be made to existing literatures but also important enough in the overall historical narrative that, when considered together, they present a new perspective on the long-run history of American space exploration.

The first of these subjects is the development of astronomical observatories—the first instruments of American space exploration—from the beginning of the nineteenth century to the middle of the twentieth. This subject, as with the broader history of American astronomy, is a critical part of the story, and its integration in economic terms provides empirical meaning to the concept of the Long Space Age. In the first half of the nineteenth century, American astronomical observatories were instruments for the personal exploration of the planets and the stars and were as well monuments of civic development. Their value was often more symbolic than scientific, and they represented significant expenditures for the individuals and communities that undertook them. Their costs were equivalent, in modern terms, to small robotic NASA probes. The cost of these facilities grew in the late nineteenth and early twentieth centuries, with the Lick, Mount Wilson, and Palomar Observatories representing major, billion-dollar equivalent investments in space exploration capabilities. These early American observatories were predominantly privately funded. Of the over forty observatories investigated, only two were built with significant government support. The motivations that dominated the financing of these "light-houses of the sky" were personal ones: intrinsic interest in the heavens and scientific curiosity, or the desire to signal status through monuments and legacies. This earliest period of American space exploration was thus one with an overridingly private context, with social entrepreneurs like Ormsby MacKnight Mitchel and George Ellery Hale selling the mystique and adventure of the heavens to the wealthy elite and the general public.

The second subject, Robert Goddard, is a case study for the interplay between private and public fund-raising and between spaceflight for scientific exploration and national defense. The American "father of liquid-fuel rocketry," Goddard had a career that constituted the world's first spaceflight development program. This book will provide an economic analysis of Goddard's life and I hope provide new insight into Goddard's motivations and financial strategy for long-run space development. Goddard was not only the first to achieve flight with a liquid-fuel rocket, he was the first to earn significant funding for spaceflight research. As with the earliest American astronomical observatories, the majority of his over $70 million (2015 GDP-ratio equivalent) funding was raised from private and semipublic sources—from the philanthropic funds provided by James Smithson, Thomas Hodgkins, Frederick Cottrell, Andrew Carnegie, Daniel Guggenheim, and Harry Guggenheim. Goddard's largest single source of support was the result of the patronage of individuals—Harry Guggenheim and Charles Abbot—who shared in his vision of a future of space travel. Though it was largely private funding that allowed Goddard to make the substantial progress that he did, he also believed that the resources required to develop the first orbital launch vehicles would vastly exceed what private individuals were likely to provide. As a result, Goddard shared with his contemporaries, Wernher von Braun in Germany and Sergei Korolev in Russia, a belief that military funding provided the key to the development of spaceflight, and he enthusiastically and persistently pursued U.S. military funding throughout his life. He was so committed to transacting this Faustian bargain that he would work with the Chemical Warfare Service on gas warfare applications and would later leave the security and long-standing patronage of the Guggenheim family in pursuit of a major military rocket development contract in the Second World War. An economic analysis of Goddard's career thus situates the motivating force of space history at the level of the individual, with the spaceflight developer willing and able to manipulate external demands, particularly in the military, in order to achieve his interplanetary objectives.

The third subject is the recent Space Age itself (circa 1950 onward). Unlike the earlier privately led phase of space exploration, it was the large-scale political demand for spaceflight that provided the driving economic force of the Cold War space race. The political history of this period has dominated the history of spaceflight and has given it an overwhelming governmental and public-sector focus, relegating the earlier history of private-sector support to the footnotes and sidelines. However, as with the largest privately funded observatories in the nineteenth century, the driving motivation for the provision of funds was a desire to signal status and capability through monumental achievement—this time at a national scale rather than the city or individual level at which earlier space exploration projects had operated. This vantage of signaling status and capability places pursuits such as the space shuttle, Space Station Freedom, and the International Space Station in the same tradition that dictated America's desire to go to the Moon.

Many historical treatments of space history have touched only incidentally on cost, but this book makes use of a quantitative estimate of space exploration expenditures over time. One constraining factor has been the limitations of the data. To conduct my analysis, I have created a new time-series of expenditures for early American astronomical observatories and for the total funds raised by Robert Goddard. Conversion of these data into 2015 gross domestic product (GDP)–ratio equivalent values and 2015 production worker compensation (PWC)–ratio equivalent values allows for comparison with modern-day space exploration efforts, in terms of their share of total American economic activities and the relative cost of their principal input—skilled labor. The results demonstrate that space exploration has mobilized major societal resource commitments in America not for decades but for centuries; that private funding has been a principal driver for the sector; and that funding for space exploration projects equivalent in cost to that of small unmanned spacecraft has been relatively common throughout the Long Space Age.

If data driven, many of this book's arguments also have conceptual roots, aimed specifically at evaluating the underlying motivations that have driven the pursuit of and allocation of resources for space exploration over the period of the Long Space Age. Particularly, I focus on two concepts:

"signaling" and "intrinsic motivation." In economic terms, these can be seen as representing, respectively, elements of the demand and the supply sides of space exploration.

The signaling value of exploration, including space exploration, has deep roots in human evolution, biology, psychology, and culture. Signaling theory also developed independently in the field of evolutionary biology. In the natural world, spectacular plumage, antlers, and other energy-expensive displays have been evolutionarily selected as "signals" of health and good genetics.[3] Less healthy individuals do not have the extra resources to devote to these nonessentials. Similarly, since genuine exploration is often a risky endeavor, the willingness to engage in it, and the capability to survive it, may serve a role in credibly conveying information about an individual or group's health and resourcefulness.

Thorstein Veblen extended signaling theory to the social phenomenon of "conspicuous consumption." The consumption of leisure, and later the consumption of costly goods, become common measures of worth and proxy signals for social capability more generally.[4] As individuals compete in ever-larger social spheres, "the means of communication and the mobility of the population now expose the individual to the observations of many persons who have no other means of judging his reputability than the display of goods."[5] Conspicuous, as in highly visible, consumption therefore becomes the way in which individuals announce their wealth and their social status to the world. As a highly visible, expensive luxury activity, space exploration can be understood as a form of conspicuous consumption—for nation states as well as for individuals—not simply because it lends a nebulous sense of prestige or pride, but because it fulfills a communication function with regard to status and capability. This notion of the credible transmission of information from one party to another through costly action is the essence of signaling.

The modern economic conception of signaling dates to the development of the economics of information in the 1970s, and it emphasized the role of signals to overcome problems of asymmetric information—a characteristic of signaling theory that is particularly important when considering the Cold War context of the Apollo program. Signals are defined in this literature as actions in which agents can choose to invest—with time, money, or other resources—in order to differentiate themselves from others. As expensive investments that result in strong reputational differentiation, space exploration activities are classic signals.

The concept of signaling is a useful tool for analyzing historical resource-allocation decisions related to space exploration. Space exploration projects are certainly costly enough to send a powerful signal. Indeed, they are good matches for Avner Offer's definition of a good signal—they are "difficult to make and difficult to fake."[6] The signaling value of space exploration projects has been a major determinant of funding decisions by national governments, institutions, and individual sponsors of space-related activity—from individual astronomical observatories to multinational space station platforms. Moreover, the signaling value of space endeavors is not merely supplemental but can often outweigh all other considerations, including those related to the advancement of science or military potential. Signaling motivations can be seen across a wide spectrum and a long time frame of space history, and they reached their peak in the Cold War space race when the signal was at its most costly, the stakes involved were at their highest, and when information asymmetries meant that space-related signals had particularly high political value.

Thus, in addition to articulating a specific, robust source of value for space exploration, the application of a signaling framework also helps to highlight the uniqueness of the Cold War circumstances that led to the space race; as spaceflight's signaling value has decreased with its relative cost and with the rise of the information age, its political relevance has diminished as well. Within this framework, the Apollo program stands out as something of an anomaly—a product of a short-term confluence of events that made it extraordinarily valuable at a geopolitical level—and thus a poor model for considering how further explorations can be conducted on a more regular and sustainable basis. The strength of its signal, however, has resulted in the story of the Apollo program continuing to shape and misshape public perceptions and American space policy for decades—long after the underlying conditions that produced that signal have changed. Signaling theory thus presents a useful tool not only for analyzing the history of space exploration expenditures and its motivational drivers but also for understanding the source of—and problems with—some of the dominant cognitive biases of the post-Apollo American space policy community.

Although an important factor, understanding the driving role of the signaling value of space exploration is thus not sufficient for understanding space exploration's long-run historical evolution, nor for considering its potential future paths. We also need to marry the influence and importance of the signaling motivations for space exploration that often drive expenditures, with the intrinsic motivations for space exploration that often drive individuals to conceptualize and pursue space exploration projects in the first place.

In its simplest formulation, "intrinsic motivation" has been defined as what people will do without external inducement.[7] It refers to behavior driven by internal interest and enjoyment that is sustainable without regard to external incentive or reward. Abraham Maslow's 1943 paper "A Theory of Human Motivation" and his 1954 book *Motivation and Personality* discussed human motivations in a framework that ran from satisfying the basic needs that people had—such as for food, safety, or love—through to the self-actualization of the individual, which he described as an impulse to be true to one's own nature.[8] It is at this level that intrinsic motivation comes into play as a driver based on an underlying need for fulfillment independent of other needs. Maslow studied virtuosos and exemplary individuals in history and concluded that some individuals were driven by a metamotivation that spurred them onward in a quest of constant improvement defined in relation to their specific underlying interests.

While seeking to avoid the "personality" trap that has swallowed up some historians of American space exploration, this book places the concept of intrinsic motivation at the core of an economic analysis. The journey into space has been a journey of self-actualization for the individuals involved—some would argue for humankind as a whole—and one in which motivations have often been divorced from immediate pecuniary returns. This journey has been driven by individuals following their intrinsic motivations and interests, devoting immense effort without regard to personal reward, and reveling in the sense of adventure and challenge. The Nobel Prize–winning chemist Harold Urey provided a charming description in 1963:

> Some 5 or 6 years ago I was interviewed by a reporter for one of the newspapers in Chicago in regard to the proposals that were

being made at the time to explore space and especially to land a man on the moon. My interview was an exceedingly discouraging one because I was not at all enthusiastic about the plans. I felt that the expense of the program would be all out of proportion to the scientific knowledge to be gained. The next morning I called up the reporter and asked that the interview not be published—in fact, that it be destroyed. The reason for the change in point of view was that overnight it had occurred to me that when men are able to do a striking bit of discovery, such as going above the atmosphere of the Earth and on to the Moon, men somewhere would do this regardless of whether I thought that it was a sensible idea or not. All of history shows that men have this characteristic.[9]

Urey is invoking a fundamental element of the socioeconomic processes underlying space exploration. There is an intrinsic human interest in exploration, and, if it is possible to explore space, then some individuals will choose to do so. Related intrinsic interests—such as the pursuit of scientific and engineering challenges, the pursuit of a multiplanetary future for humanity, or the simple pursuit of personal adventure in space—factor into the development of space exploration in a similar manner. Those individuals—and most specifically the exemplary virtuosos—that choose to dedicate their efforts to the exploration of space, regardless of remuneration, are thus a leading input of the production function of spaceflight.

Urey's pithy anecdote leaves out an important factor, however—that there must be two sides to the economic equation if the expensive instruments of space exploration are to be produced. The intrinsic interests of individuals must be matched by a perception that the activity is worthy of resource allocation. Parties entering into an economic exchange around space exploration can share the same intrinsic motivations, or they can differ substantially, but there must be commitment on both the supply and demand side. As indicated earlier, often it is the demand for monuments and signaling that has provided the resources to those intrinsically motivated by space exploration to fulfill their ambitions. Sometimes, however, resources are allocated to a space exploration project by an intrinsically motivated patron. At other times, an individual self-supplies the

funding for his or her own space exploration projects—making it particularly difficult to differentiate the extent to which this decision was intrinsically motivated and the extent to which it was driven by a desire to
signal. This complex decision-making interplay of intrinsic motivations
and signaling is at the core of what we are setting out to examine in a
wide-ranging set of astronomical observatories and spaceflight technology development programs.

As we examine the motivations for and the circumstances of decision-
making on American space exploration projects, there is one element of
those circumstances that will be repeatedly noted and discussed at length:
the source of funding. This investigation categorizes projects and programs into one of two principal sources of funding: public sources, allocated according to the public good (as determined, in this case, by the
American political system) and provided by taxation; and private sources,
allocated according to the interests and desires of the individuals providing the resources.

Why is the determination of whether a project was publicly funded or
privately funded so important? Why categorize in terms of public or private,
rather than in terms of governmental or commercial funding, or another
set of characteristics? Although we have not seen a shift in the *motivations* for space exploration over the past two hundred years, we have seen
such a shift in the dominant *source of funding* for American space exploration over that period. Measuring, analyzing, and articulating that shift
are primary objectives of this book. Investigating the funding of past space
exploration raises important social, ethical, and political questions, not
all of which we can investigate here. Instead, my hope is that this analysis
will give readers a better understanding of how the current "privatization
of spaceflight" fits into a much longer history. Perhaps more importantly
it will also allow readers to consider how the long-run forces described
might drive the next great explorations of the heavens.

1

PIETY, PIONEERS, AND PATRIOTS: THE FIRST AMERICAN OBSERVATORIES

Each expedition into remoter space has made new discoveries and brought back permanent additions to our knowledge of the heavens. The latest explorers have worked beyond the boundaries of the Milky Way in the realm of spiral "island universes," the first of which lies a million light-years from the earth while the farthest is immeasurably remote.

—*George Ellery Hale, "The Possibilities of Large Telescopes," 1928*

Long before rockets allowed us to explore the solar system, humans explored space through another expensive technology—large telescopes. The American astronomical observatories of the nineteenth and early twentieth centuries were projects of considerable complexity. From these institutions, astronomers would embark on their "exploration of the heavens," as their activities were regularly referred to, and convey their findings to a public eager for new discoveries. Early American observatories employed new technologies—requiring expensive imports and engineering contracts—and mobilized significant capital expenditures. Observatories also often grew to possess a political and cultural importance that outweighed their scientific significance, becoming objects of community pride and signals of national development. And yet despite all this—and the striking similarities of these projects to modern missions of space exploration—the economic history of American astronomical observatories has not been integrated into the broader narrative of American space history.

Telescopes and robotic probes launched by rockets are very different technologies. In one respect, however, they are very similar: the experience of the human observer, whether using a telescope or spacecraft to explore space, is fundamentally the same—that of using technology to extend vision into space. By using a consistent metric to compare the cost of that technology, whether spacecraft or telescope, we can examine the funding in America for the exploration of the heavens as a continuum from the mid-nineteenth century to the present day and identify long-run trends.

There are a number of metrics that could be used to compare historical costs and convert them into present-day equivalent values.[1] The most commonly used method is adjustment for inflation using a consumer price index. This method calculates current equivalent values of historical costs in terms of the change in the price of an index of basic consumer goods, like bread and clothing, over the intervening time. Although this is a useful metric for evaluating the equivalent buying power of historical values in terms of basic consumer goods, it is not a very useful metric for evaluating the economic significance of large capital expenditures within the context of the overall economy. This metric has been the dominant method of conversion in the history of astronomy literature, with the result that the modern equivalent costs that have been suggested in the historiography have often been significantly understated.[2]

There are ways of converting historical observatory expenditures into modern contexts that are more meaningful—a conversion in terms of the cost of production worker compensation (PWC) for a given project, and in terms of the ratio of the Gross Domestic Product (GDP) of the American economy that a project represented. The appropriateness of the metric is determined by the question that is being asked. If we are interested in how much it would cost to build the same observatory today, then we should adjust the expenditure in terms of the principal cost of production—the labor costs of the manufacturing production workers who built the observatory. An expenditure amount adjusted by the change in production worker compensation over the intervening time—the PWC metric—is a useful measure for comparing the overall complexity of the project in modern terms, as the principal cost of a modern project of space exploration is also the labor cost of its manufacturing production

workers. Alternately, if we are interested in the share of total economic resources that a given project represents, we need to examine the project
expenditure relative to the size of the U.S. economy as a whole. To do
this, we take the expenditure of a given project and divide it by the estimated GDP of the U.S. economy for the year the funds were committed.
To convert this ratio into an equivalent present-day value for that project,
the percentage is multiplied by the current GDP. The result is the equivalent GDP-ratio value of the project expenditures. If we are interested in
how the American economy has allocated its resources toward the exploration of the heavens over time, this is the appropriate metric to examine. It tells us, for any historical value, how much one would need today
to undertake a project that would be equivalent in its share of the resources of the American economy to what the historical project represented in its day.

The observatories examined in this chapter are listed in table 1.1, along
with their costs and calculated equivalent PWC- and GDP-ratio values.[3]
The sources for the GDP data are from Johnston and Williamson, and
the sources for the production-worker-compensation data are from Officer.[4] The list only includes those observatories for which relatively robust
estimates of total expenditure can be determined, which leaves out a handful of the observatories discussed below—most notably, the observatories
of the University of Alabama, the University of Mississippi, the Lowe Observatory, and the observatories of Rutherford, Draper, and Lowell. Plots
of the equivalent PWC- and GDP-ratio adjusted equivalent values for the
different observatories over time can be seen in figures 1.1 and 1.2, while
histograms of decadal expenditures using the same metrics can be seen
in figures 1.3 and 1.4. Summary statistics are in table 1.2.

Examining the 2015 GDP-ratio equivalent values of early American observatories makes it clear that projects aimed at the exploration of the
heavens have involved economically significant levels of American resources
for over 150 years. Projects equivalent to $100 million to $1 billion were
relatively common (see figure 1.1). Many of the observatories are of an
equivalent relative magnitude to major modern ground-based observatories, such as the Gemini Observatories, the Grand Canary Telescope, and
the two Keck Telescopes, or to early NASA Discovery–class robotic interplanetary missions (all approximately $200 million–$250 million).[5] The

Table 1.1. Expenditures on U.S. Observatories, 1820–1940

Project	Year*	Nominal prices in U.S. dollars ($)	Constant prices in U.S. dollars ($), adjusted by PWC index, base year 2015	GDP-ratio equivalent value in U.S. dollars ($), adjusted by ratio to GDP, base year 2015
Yale College Observatory	1828	1,200	764,000	24,100,000
University of North Carolina Observatory	1831	6,400	3,490,000	110,000,000
Hopkins Observatory	1836	6,100	3,580,000	74,400,000
Hudson Observatory	1836	4,000	2,350,000	48,800,000
Philadelphia High School Observatory	1837	5,000	2,500,000	58,000,000
West Point Academy	1842	5,000	2,390,000	55,700,000
U.S. Naval Observatory	1842	25,000	11,900,000	279,000,000
Cincinnati Observatory	1843	16,000	8,730,000	184,000,000
Harvard College Observatory	1844	50,000	26,800,000	530,000,000
Georgetown Observatory	1844	27,000	14,500,000	286,000,000
Jackson Observatory	1845	4,000	2,140,000	38,800,000
Edward Phillips Endowment—Harvard	1848	100,000	47,000,000	743,000,000
Shelby College Observatory	1848	3,500	1,640,000	26,000,000
Detroit Observatory	1852	22,000	10,000,000	129,000,000
Dartmouth College Observatory	1852	11,000	5,020,000	64,700,000
Litchfield Observatory	1854	50,000	22,500,000	243,000,000
Dudley Observatory	1854	119,000	53,500,000	578,000,000
Allegheny Observatory	1862	32,000	10,700,000	98,800,000
Vassar College Observatory	1865	14,000	3,820,000	25,300,000
Dearborn Observatory	1865	56,000	15,300,000	101,000,000
Winchester Observatory	1871	100,000	26,300,000	235,000,000
Halsted Observatory	1872	60,000	15,700,000	130,000,000
Morrison Observatory	1874	100,000	25,900,000	211,000,000
Lick Observatory	1876	700,000	188,000,000	1,510,000,000
Washburn Observatory	1876	65,000	17,400,000	140,000,000
Warner Observatory	1880	100,000	27,500,000	172,000,000
McCormick Observatory	1881	135,000	37,500,000	207,000,000

(continued)

Table 1.1. (*continued*)

Project	Year*	Nominal prices in U.S. dollars ($)	Constant prices in U.S. dollars ($), adjusted by PWC index, base year 2015	GDP-ratio equivalent value in U.S. dollars ($), adjusted by ratio to GDP, base year 2015
Kenwood Physical Observatory	1888	25,000	5,970,000	32,200,000
Elias Loomis Endowment—Yale College	1889	300,000	68,900,000	387,000,000
Goodsell Observatory	1890	65,000	14,900,000	77,000,000
Chamberlin Observatory	1890	56,000	12,900,000	66,400,000
Ladd Observatory	1891	30,000	6,890,000	34,800,000
Yerkes Observatory	1895	349,000	84,600,000	400,000,000
McMillan Observatory	1895	16,000	3,880,000	18,300,000
New Allegheny Observatory	1906	300,000	56,200,000	173,000,000
Mount Wilson Observatory	1910	1,450,000	253,000,000	775,000,000
Griffith Observatory	1919	225,000	15,900,000	51,300,000
Perkins Observatory	1925	379,000	23,000,000	74,800,000
Palomar Observatory	1928	6,550,000	386,000,000	1,200,000,000
McDonald Observatory	1929	840,000	49,700,000	145,000,000

*Base year used for calculations; see endnote 3 to chapter 1.
Sources: See text of chapters 1 and 2 for observatory cost references. PWC-ratio equivalent value and GDP-ratio equivalent value calculations done using measuringworth.com.

sum total of the expenditures made on American astronomical observatories in the data set is nearly $10 billion in 2015 GDP-ratio equivalent terms. For ease of reading, henceforth 2015 GDP- and PWC-ratio equivalent terms will be shortened to GDP-ratio terms and PWC-ratio terms accordingly.

A number of the nineteenth-century observatories, such as the Lick Observatory and the Palomar Observatory, were equivalent to major NASA missions, such as the New Horizons mission to Pluto ($670 million), the

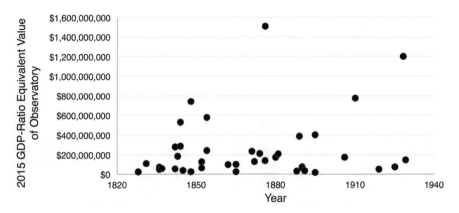

Fig. 1.1. Expenditure on U.S. observatories, 1820–1940: GDP-ratio adjusted equivalent value in 2015 U.S. dollars. (Source: Compiled by author from data referenced in text and notes.)

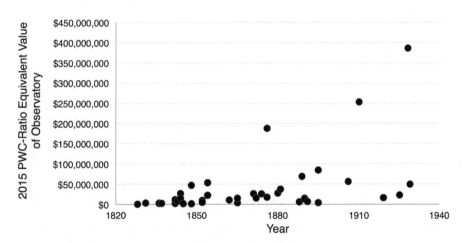

Fig. 1.2. Expenditure on U.S. observatories, 1820–1940: constant prices in 2015 U.S. dollars—PWC-ratio adjusted equivalent value. (Source: Compiled by author from data referenced in text and notes.)

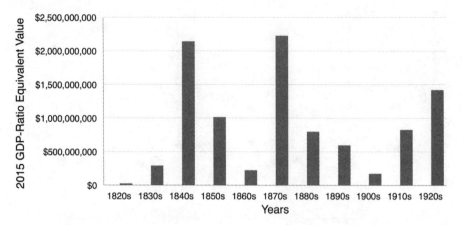

Fig. 1.3. Decadal expenditure on U.S. observatories, 1820s–1920s: GDP-ratio adjusted equivalent value in 2015 U.S. dollars. (Source: Compiled by author from data referenced in text and notes.)

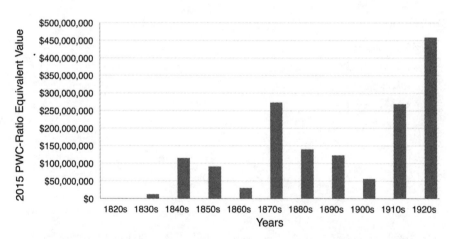

Fig. 1.4. Decadal expenditure on U.S. observatories, 1820s–1920s: constant prices in 2015 U.S. dollars—PWC-ratio adjusted equivalent value. (Source: Compiled by author from data referenced in text and notes.)

Table 1.2. Expenditures on U.S. Observatories, 1820–1940, Summary Statistics

Total number of observatories and endowments in data set	40
Total PWC-ratio adjusted value of expenditures in 2015 U.S. dollars	$1,568,764,000
Total GDP-ratio adjusted value of expenditures in 2015 U.S. dollars	$9,737,400,000
Percentage of total GDP-ratio equivalent expenditures supplied by government funds	3.4%
Percentage of total GDP-ratio equivalent expenditures supplied by private-sector funds	96.6%

Source: Table figures compiled by author.

MESSENGER probe to Mercury ($420 million) or the Mars Exploration Rovers ($850 million).[6] The modern equivalent values calculated here should be taken only to indicate the relative order of magnitude of expenditure, given that other appropriate equivalent resource share values can be calculated. For example, rather than scaling the expenditure as a share of the total resources of the U.S. economy, the expenditure can be scaled as a share of the resources of the individuals who undertook the projects. James Lick was the richest man in California, and the Lick Observatory expenditure represented 17.5 percent of his entire estate. The equivalent share of the wealth of the richest man in California in 2015, Larry Ellison, was roughly $9.5 billion, approximately six times higher than the GDP-ratio equivalent share. The order of magnitude calculations are clear: there were well over a dozen astronomical observatories built in America in the nineteenth and early twentieth centuries that were of comparable relative economic significance to modern robotic spacecraft. The continuum of this major funding flow in the history of American space exploration has been significantly underappreciated.

Although the GDP-ratio adjusted values are the most striking, the PWC-ratio adjusted constant values also show that the projects were approaching the technical complexity of modern space missions (see figure 1.2). Over twenty of the observatories were equivalent in these terms to small spacecraft projects ($10 million–$100 million) while three—the Lick, Mount Wilson, and Palomar Observatories—were equivalent to full-scale NASA planetary science missions, like the $224 million NEAR Shoemaker mission to the asteroid 433 Eros.[7] Although this comparison relates the most technically complex of the observatories in the study with midclass robotic space exploration missions, there is also almost three-quarters of a century of economic growth and technology development between them. As we would expect in any exploratory process where it becomes increasingly difficult and costly to conduct new exploration as the lower-hanging fruit is progressively picked off, the top flagship projects for the exploration of the heavens become more technically complex over time. This trend continues well into the Space Age: the Apollo program cost some $25 billion in 1969 dollars, equivalent to roughly $205 billion in PWC terms. That the technical complexity of modern space exploration projects is significantly greater than most of the early American observatories is not at all surprising. What is surprising is that a handful of the early observatories *are* comparable to the technical complexity of modern spacecraft and, perhaps even more importantly, that these complex and expensive projects were almost entirely privately funded.

The dominance and economic significance of the private funding of American projects for the exploration of the heavens in the nineteenth and early twentieth centuries, as shown in table 1.2, is a significant finding. Although the fact that private funding was an important factor in the development of American astronomy has been shown by Howard Miller, and has been shown more generally for nineteenth-century science and technology projects by Terence Kealey, the full extent of this private funding for American astronomical observatories has been seldom appreciated in the context of space history.[8] To put the extent of the private support for astronomy within context, of the thirty-eight observatories listed, only two—the U.S. Naval Observatory and the Observatory of the U.S. Military Academy at West Point—were not privately owned observatories with large optical telescopes. In pursuing personal passions and public monuments, American citizens, through collective subscription campaigns and singular philanthropy, privately funded the increasingly expensive technology required for the continued exploration of the heavens for over a century before NASA or the invention of the liquid-fuel rocket.

In the earliest period of American history, astronomy and the exploration of the heavens were considered a hallmark of intellectual development and

a noble endeavor for the colonial elite. John Winthrop Jr., eldest son of
the first governor of the Massachusetts Bay Colony, and later governor
himself, had an intense personal interest in astronomy, which led him to
become the first American member of the Royal Society, to correspond
with Sir Isaac Newton, and to import the first telescope to the New World
in 1660.[9] Colonial interest in astronomy was more pronounced than in any
nonagricultural science, due in no small part to the inclination of the Puri-
tan clergy in New England to regard the subject as manifesting the work
of God. This would be a recurring theme and play a significant role in motivating the establishment and financing of early American observatories.
A collective religious fervor supported a community interest in studying
"the heavens," although scientific astronomical observations were made
by individuals in isolation, as a networked scientific community was
largely precluded by the difficulties of communication across the American wilderness; in fact, colonists often relied upon English correspondents for news of other colonists.

By the early eighteenth century, a handful of individuals had begun to
accumulate significant astronomical resources, and astronomy had become
a relatively common theme in American intellectual society. The science
was driven forward largely by the passion of a few notable virtuosos, such
as James Logan and David Rittenhouse, whose efforts and enthusiasm
contributed to the development of Philadelphia as the scientific capital of
colonial America. Logan assembled the finest scientific library in the colonies, with books by Copernicus, Galileo, Brahe, Kepler, Huygens, Halley,
Flamsteed, Hevelius, and Ptolemy, and three editions of Newton's *Principia*
at a time when Harvard had none.[10] By midcentury a number of major
universities had at least a part-time astronomer. Astronomy was a sufficiently prestigious pursuit that it was a common study of college presidents. Yale's Ezra Stiles and Harvard's Joseph Willard—who corresponded
with Nevil Maskelyne, the Astronomer Royal—were both practicing observers.[11] Although the contribution of the colonies to the astronomical
literature was limited, a general knowledge of astronomy was a sign of re-
finement in society and an interest of a number of the early colonial leaders. The June 8, 1774, diary entry of young Philip Vickers Fithian, tutor
to the children of Virginia planter, government official, and gentleman
Colonel Robert Carter (from one of the most famous of the Virginia

families), is an excellent example of the fluid, informal way in which an interest in astronomy and the solar system was mixed into the intellectual ferment of the colonial era elite: "The morning pleasant—Mr. Carter rode to the Ucomiko Warehouses to examine in the Shipping some of his Tobacco—We have no Company. The day is very warm—A flaming sultry Sun, a dusty scorched Ground, Mr. Carter returned, the day being smoky introduced, at Coffee, a conversation on Philosophy, on Eclipses; the manner of viewing them; Thence to Telescopes, & the information which they afforded us of the Solar System; Whether the planets be actually inhabited &c."[12]

With the observations of the transits of Venus—rare crossings of the planet Venus passing in front of the face of the Sun from Earth's perspective—the study of the heavens in British North America also began to assume a political context. Their careful observation can enable the derivation of the solar parallax and thereby the distance from the Earth to the Sun and the size of the solar system. Throughout the eighteenth and most of the nineteenth centuries, this measurement was considered to be a "holy grail" of science, holding out the prospect of great prestige for any nation or individual associated with its accurate attainment. The transits thus became major events in international politics as well as in science. For the 1761 transit, for instance, the czarina of Russia, the kings of France and England, and the directors of the East India Company all sent expeditions to observe the transit. Eager to show willingness and ability to participate in the same grand endeavors as the Old World, the Province of Massachusetts Bay outfitted a sizable expedition to St. John's, Newfoundland, where Harvard Professor John Winthrop and his assistants set up a small observation camp.[13] Winthrop's observations were the only American input incorporated into the 1761 transit calculations and generated little international or even American colonial interest. The American observation of the 1769 transit would be another matter, however, as it would provide extensive sighting opportunities from within the borders of the American colonies.

For the 1769 transit, the American Philosophical Society, with a telescope from the Penn family and financial support of around £200 from the Pennsylvania Provincial Assembly, organized observers in Philadelphia and Norriton, Pennsylvania, as well as in Cape Henlopen, Delaware.[14]

Separate observations were arranged in the colonies of Rhode Island, Massachusetts Bay, and New Jersey and in Canada.[15] The results of the observations were widely regarded as the highlight of the first issue of the *Transactions of the American Philosophical Society*. The quantity and quality of the American observations impressed many in Europe, and they were interpreted as a signal of a "new stage of maturity in the development of America."[16] On the whole, the effort was deemed a significant symbol of national accomplishment and pride and when the Declaration of Independence was first read publicly on July 8, 1776, to the mass of people assembled in State House Square of Philadelphia, it was from the platform of the temporary observatory built for the transit.

In the wake of the American Revolution, the desire to signal a robust and independent national presence was intensified in all areas, including astronomy. In 1781, the same year that the British Army surrendered at Yorktown, the Pennsylvania legislature provided a grant to establish the first American observatory of any permanence, that of David Rittenhouse in Philadelphia, which remained in some form of operation until 1810.[17] Rittenhouse was the most accomplished colonial astronomer, contributing significantly to the 1769 transit of Venus observations and constructing a highly accomplished orrery—a mechanical model of the solar system. Rittenhouse's orrery was the most exact and precise of his day, and it made him a cultural hero to the educated men of the Revolution. In 1771, the Pennsylvania Assembly granted him £300 for the achievement, and the presidents of the College of New Jersey and the College of Philadelphia paid £300 pounds each for copies of the device.[18] For comparison, Harvard College had paid £92 10s 6d for an orrery from the famed London instrument maker Benjamin Martin in 1767.[19] The Pennsylvania Assembly offered Rittenhouse an additional £400 for a copy intended for public use. In response to critics of the newly independent America, Thomas Jefferson put forth the self-taught Rittenhouse (along with George Washington and Benjamin Franklin) as one of the three great intellects who signaled the young nation's coming of age: "We have supposed Mr. Rittenhouse second to no astronomer living; that in genius he must be the first, because he is self taught. As an artist he has exhibited as great a proof of mechanical genius as the world has ever produced. He has not indeed made a world; but he has by imitation approached nearer its Maker than any man who has lived from the creation to this day."[20] Rittenhouse was even nominated for a newly envisioned position, the "Astronomer to the State of Pennsylvania," with a salary of £500, more than twice the salary received by William Herschel, the discoverer of the planet Uranus, from King George III.[21] Although the position was never actually established by the assembly, it shows the regard that some felt was owed to Rittenhouse for the work he had rendered in the service of science and, thereby, in the service of the new nation's standing in the world. The assembly did eventually grant Rittenhouse £250 in 1781 for the construction of an observatory of his own in Philadelphia.[22] As with the transit of Venus observations, the achievements of David Rittenhouse are early and important examples of the importance placed by men of influence in colonial America on projects of astronomy as signals of national accomplishment.

Consistent with this sentiment, there were repeated attempts to establish a more permanent American observatory in the early nineteenth century. In Washington, Ferdinand Rudolph Hassler, director of the U.S. Coastal Survey, proposed an observatory to assist with the tasks associated with a surveying project launched in 1807. William Lambert, congressional clerk and scribe of the Bill of Rights, presented a memorial to Congress in 1809 advocating for an observatory to establish an independent Washington meridian for the development of a national standard of longitude—a proposal that received the support of a congressional committee and Secretary of State James Monroe but for which no funds were ultimately appropriated.[23] Similarly unsuccessful was the American Philosophical Society's appeal to the Pennsylvania General Assembly for funds to convert an abandoned pumping station into an observatory in 1817.[24] In 1820, Thomas Jefferson planned to create a large observatory at the University of Virginia, which he estimated would cost $10,000–$12,000 ($6.9 million–$8.3 million in PWC-ratio terms; $254 million–$305 million in GDP-ratio terms), roughly 6–7.5 percent of the estimated cost of the entire university, as well as a planetarium for the university's rotunda dome.[25] He only managed, however, to construct a rudimentary building for observations, one that was never used and later fell into ruin. Harvard attempted to organize an observatory initiative no less than five times, in 1805, 1816, 1822, 1823, and 1825, but the public subscription efforts failed to raise a sufficient sum.[26] Harvard's attempts, however, drew support and a pledge of $1,000 of personal funds from the man who would become the first great advocate for U.S. astronomy and, I would argue, for the American exploration of space: the sixth American president, John Quincy Adams.

John Quincy Adams was an intellectual and worldly president. His early years had been spent largely in Europe, where his father John Adams, the second U.S. president and leading orator of American independence, had served as American envoy to France, the Netherlands, and Great Britain. John Quincy Adams's astronomical interest may well have been kindled during his studies at Leiden University, a renowned institute of science with a venerable university observatory dating back to 1633. He arrived in the presidency conscious of European astronomical achievements and with a determination to ensure that the new American nation would be able to match them. As secretary of state he had been a strong advocate of "continentalism," the belief that the United States had a destiny to occupy the North American continent. He had supported expansionist advances and was also the author of the 1823 Monroe Doctrine, which put a shot across the bows of any additional European colonial ambitions in the Americas. He carried this spirit of competitive American patriotism into his advocacy for astronomical endeavor. In his first annual message as president in 1825, he urged the establishment of an American astronomical observatory in connection with a proposed national university and, in doing so, evoked themes of national boosterism and technological competition, which were echoed in the space race nearly 150 years later:

> Connected with the establishment of a university, or separate from it, might be undertaken the erection of an astronomical observatory, with provision for the support of an astronomer, to be in constant attendance of observation upon the phenomena of the heavens; and for the periodical publication of his observations. It is with no feeling of pride, as an American, that the remark may be made that, on the comparatively small territorial surface of Europe, there are existing upward of one hundred and thirty of these light-houses of the skies; while throughout the whole

American hemisphere there is not one. If we reflect a moment upon the discoveries which, in the last four centuries, have been made in the physical constitution of the universe by the means of these buildings, and of observers stationed in them, shall we doubt of their usefulness to every nation? And while scarcely a year passes over our heads without bringing some new astronomical discovery to light, which we must fain receive at second-hand from Europe, are we not cutting ourselves off from the means of returning light for light, while we have neither observatory nor observer upon our half of the globe, and the earth revolves in perpetual darkness to our unsearching eyes?[27]

It is notable that science was far from central in the argument. The focus was instead on how America lagged behind Europe in the field. Although his writings and orations show John Quincy Adams to have been a firm believer in the importance of science and exploration for their own sake, he chose to tailor his arguments for observatory funding to appeal to American pride and prestige.

Adams's observatory plan was unprecedented in its ambition as a federal government enterprise. Following on his impassioned inaugural speech, his official proposal came forward in 1826, requesting an initial outlay of about $15,000 and an annual operating cost of $4,000 ($9 million and $2.4 million respectively in PWC-ratio terms; $312 million and $83 million in GDP-ratio terms).[28] Large government projects, such as canals, roads, and surveys, were relatively common at the time, and much more costly—the Erie Canal, a state-level initiative, was estimated at $7.14 million when completed in 1825, while the federal National Road project that connected towns from Cumberland, Maryland, to Vandalia, Illinois, totaled some $6.8 million in appropriations between 1811 and 1838.[29] However, no scientific or institutional project of parallel magnitude had yet been undertaken. Not surprisingly, therefore, he faced strong, even vitriolic, opposition to the idea. The states saw such a federal project as encroaching on their jurisdictional responsibility over education, which was seen as intertwined with astronomy. These concerns were not entirely ill placed, as Adams was explicitly trying to use astronomy and the observatory to create a potent national symbol of a unified America. The newly independent, and only loosely federated states had little interest in seeing any such symbol established with their funds. They greeted the proposal with such ridicule and opposition that, as Loomis reported in 1856, the president's phrase "light-houses of the skies," often misquoted by his opponents as "light-houses *in* the skies," became associated with spurious, wasteful ideas to such an extent that it was used as a political slur even into the 1850s.[30] As A. Hunter Dupree put it, "No more hated proposal existed and nowhere had more pains been taken to prevent the creation of a new agency."[31] The notion of a federal observatory had become such a polarizing issue that, in 1832, long after John Quincy Adams had left office (although he was still a member of Congress), the legislature took pains to ensure that it would not resurface. An act in that year for the continuance of the survey of the coast included the proviso that "nothing in the act should be construed to authorize the construction or maintenance of a permanent astronomical observatory."[32]

Congress was right to be concerned about the idea's resurrection. The determined Adams would try again to establish a national observatory over a decade later when a new opportunity arose with the Smithson bequest. James Smithson, an Oxford-educated mineralogist, chemist, and successful investor, had left his legacy to the United States, in spite of never having visited the country, on condition that it be used "for the increase and diffusion of knowledge among men." The bequest touched off a transoceanic court battle and a broad, years-long debate in America about the appropriate use of the funds. There was much to argue about as the bequest, when eventually accepted in 1836, amounted to $508,318.46 ($299 million in PWC-ratio terms; $6 billion in GDP-ratio terms). Although a number of small private and college observatories had been established in the interim, the country still lacked a major or national observatory. Adams seized the opportunity to relaunch his campaign for such an institution, arguing that it would be the most fitting fulfillment of Smithson's wishes.[33]

As chair of the House of Representatives Committee on the Smithsonian Bequest, Adams threw himself anew into his campaign for an astronomical observatory. The characteristic soaring Adams prose, which the father had used to rouse the Continental Congress to vote for independence, was employed by the son in praise of the importance of astronomy for the nation and for mankind in general:

The influence of the moon, of the planets—our next-door neigh-bours of the solar system—of the fixed stars scattered over the blue expanse in multitudes exceeding the power of human com-putation, and at distances of which imagination herself can form no distinct conception; the influence of all these upon the globe which we inhabit and upon the condition of man, its dying and deathless inhabitant, is great and mysterious, and, in search for final causes, to a great degree inscrutable to his finite and limited faculties. The extent to which they are discoverable is, and must remain unknown; but to the vigilance of a sleepless eye, to the toil of a tireless hand, and to the meditations of a thinking, com-bining, and analyzing mind secrets are successively revealed, not only of the deepest import to the welfare of man in his earthly career, but which seem to lift him from the earth to the thresh-old of his eternal abode; to lead him blindfold up to the Council Chamber of Omnipotence, and there stripping the bandage from his eyes, bid him look undazzled at the throne of God.[34]

Adams was not advocating for science in general, but for astronomy in particular. He was a strong supporter of science, but he evidently placed even greater value on the exploration of the heavens: "There is not one study in the whole circle of sciences more useful to the race of man upon earth or more suited to the dignity of his destination, as a being endowed with reason, and born to immortality, than the science of the stars."[35] His position seems to have stemmed partly from a belief that astronomy was a frontier of limitless potential, defining it as the "systematic and contin-ued scientific series of observations on the phenomena of the numberless worlds suspended over our heads—the sublimest of all physical sciences, and that in which the field of the future discovery is as unbounded as the universe itself."[36] For Adams, astronomy was part of a higher calling. "The science of astronomy," he wrote, "is the intercourse of immortal man with the universe."[37] This lofty prose, common throughout Adams's writing on astronomy, harkens back to the association of astronomy with religion as seen in early colonial America. Adams translates this to a more human-istic Enlightenment context, arguing that the exploration of the heavens has a special role in the destiny and purpose of humanity and that it is therefore deserving of special government support. The rhetoric even rose to identifying astronomy as the core mission of humanity: "So peculiarly adapted to the nature of man, is the study of the heavens, that of all animated nature, his bodily frame is constructed, as if the observation of the stars was the special purpose of his creation."[38] Adams placed astronomical observation into a narrative that emphasized exploration and discovery, writing, "there is no richer field of science opened to the exploration of man in search of knowledge than astronomical observation," and stressing, "the express object of an observatory is the increase of knowledge by *new discovery*."[39] This focus on astronomical discovery was somewhat novel in reference to a national observatory. The national observatories at Greenwich and Paris had served largely utilitarian functions for navigation and timekeeping, although considerable exploratory astronomy had been conducted by the observatories' astronomers once they were established. The first national observatory to be established with a direct mandate to engage in such exploration from the beginning was the Pulkovo Observatory in Russia, and it seems to have been the purpose and scale of Pulkovo that directly influenced Adams's proposal.

The Pulkovo Observatory, which had been founded outside St. Petersburg in 1839, was the brainchild of astronomer Friedrich Georg Wilhelm von Struve. Struve had convinced Tsar Nicholas I, then pursuing a suite of policies to promote Russian nationalism, to build the world's most well appointed, and thus most expensive, observatory as a symbol of Russian greatness.[40] Adams had lived in St. Petersburg from 1809 to 1814 as the first United States minister to Russia and was well versed on the details of the project. He went to pains to point out to his fellow Americans the extent of the support the observatory received:

> A plan was accordingly prepared in April, 1834, with estimates of the expense required for the erection of the building, and for the purchase of suitable instruments for observation; the first amounting to 346,500 rubles, and the second to 135,000 rubles—the value of the ruble being 74 cents of our currency. The Emperor immediately issued an order to commence the work, under the direction of three commissioners, members of the academy; and that 100,000 rubles should be placed at the disposal of

the minister of publication instruction, for the prosecution of the work. It was accordingly commenced, vigorously prosecuted, and finally completed, at a cost of little less than one million rubles.[41]

The scale of the Pulkovo Observatory led Adams to propose an observatory of similar scale for America. Based on information regarding costs for observatory construction, maintenance, salaries, and instruments received from George Airy, the Astronomer Royal at Greenwich, Adams's report proposed that the annual interest on the capital of the Smithson bequest be used to pay for an American national observatory over a course of seven years. The proposed expenditure amounted to $210,000 ($113 million in PWC-ratio terms; $2.4 billion in GDP-ratio terms), with the appendix of the report later advising that the amount be increased to ten years of the interest for a total of $300,000 ($161 million in PWC-ratio terms; $3.4 billion in GDP-ratio terms).[42] Adams's broad plan for the Smithson bequest was to use the annual interest from the principal to fund the establishment of relevant projects and institutions in perpetuity. He preferred an astronomical observatory to be the first such project, given his belief that it was the preeminent science and his conviction that the support of astronomy was an important step for the young American republic to take in its development and to mark its standing in the world.

In his rhetoric, Adams relied heavily on an appeal to American pride and specifically noted the esteem in which even autocratic Russia was held on account of its efforts in astronomy. Adams's committee report from 1840 merits quotation in length:

> Here is the sovereign of the mightiest empire and the most absolute government upon earth, ruling over a land of serfs, gathering a radiance of glory around his throne by founding and endowing the most costly and most complete establishment for astronomical observation on the face of the earth. . . . And this event is honorably noticed in the National Institute of France, one of the most learned and talented assemblies of men upon the globe noticed as an occurrence in the annals of science noticed for honor and for emulation. The journalist of a free country, applauding the exertions of a land of serfs to promote the progress of science, avows that he should blush for his own country, had he not at

hand the evidence of her exertions not less strenuous for the advancement of the same cause.

The committee of the House, in applying to their own country that sensibility to the national honor which the French journalist attributes to self-love, would gladly seek for its gratification in the same assurance that she is not lagging behind in the race of honor; but that, in casting their eyes around over the whole length and breadth of their native land, they must blush to acknowledge that not a single edifice deserving the name of an astronomical observatory is to be seen.[43]

The similarities to the rhetoric used in Congress in 1840 and in the 1950s and 1960s to bolster support for the American space exploration effort are striking. There is a basic appeal to American pride, an emphasis on the implications for international recognition and prestige, and even an explicit reference to a "race of honor." The parallel to the rhetoric of the Cold War space race is complete, with Adams situating this "race" as one between an autocratic Russian state and a proudly democratic America.

Unlike in the 1950s, however, the appeal to national pride and prestige was unsuccessful. The continued widespread opposition to Adams's conception of a large federal observatory could not be overcome. Nevertheless, Adams's vision of an institution that focused on supporting new research and discovery influenced the ultimate use of the Smithson bequest. This broad-ranging vision allowed the Smithsonian Institution to contribute to the exploration of the heavens in ways that even Adams had not foreseen. The Smithsonian became a significant early advocate and source of support for astrophysics research, with noted American astrophysicist and early flight pioneer Samuel Pierpont Langley becoming the institution's third director and establishing the Smithsonian Astrophysical Observatory in 1890 with $10,000 for solar astronomy work.[44] Perhaps even more significantly, it was the Smithsonian Institution that published American spaceflight pioneer Robert Goddard's "A Method for Reaching Extreme Altitudes" and provided him with his first significant financial support. Indeed, it is the Smithsonian Institution, shaped in the early nineteenth century by John Quincy Adams's passion for astronomy, that best embodies the continuity of U.S. space exploration efforts across the centuries and that also houses the secular temple of American space-flight, the Smithsonian National Air and Space Museum.

In addition to the legacy of the Smithsonian, Adams was also successful in raising the status and importance of astronomy in American life. He arguably did more than any of his contemporaries to generate and cultivate the demand for astronomy and the observatory-building enthusiasm that took hold throughout the United States in the mid-nineteenth century. When one of the first major American observatories was established in Cincinnati in 1843, it was in no small part due to the prominence and visibility that Adams had given to astronomical science. The seventy-seven-year-old Adams made the grueling trip from Boston to Cincinnati to deliver his last major public oration, the dedication address at the observatory atop the newly renamed "Mount Adams."

Although both his attempts to establish a national observatory failed, Adams set a rhetorical tone that would long endure in the advocacy of American space exploration. He drew heavily on the view, first espoused in America by the Puritans, that astronomy "elevates those who study it because its object is God's handiwork."[45] To Adams, the science of astronomy was a duty to the spirit of America's forefathers, to God, and, in the best Enlightenment tradition, to the fulfillment of humanity's potential. Mindful of his pragmatic Jacksonian audience, however, he also emphasized more down-to-earth benefits, principally international prestige and navigational improvement: "That by the establishment of an observatory upon the largest and most liberal scale, and providing for the publication of a yearly nautical almanac, knowledge will be diffused among men, the reputation of our country will rise to honour and reverence among the civilized Nations of the earth, and our navigators and mariners on every Ocean, be no longer dependent on English or French observers or calculators for the tables indispensable to conduct their path upon the deep."[46] The lofty personal motivations allied to practical arguments, the emphasis on national identity, and the explicit reference to international prestige would reemerge in the *Congressional Record* during discussions of the U.S. transit of Venus expeditions in the late nineteenth century and in the twentieth-century debates on the space programs.

More importantly, however, John Quincy Adams was a catalyst for the development of a popular interest in astronomy in America. The numerous failed attempts to establish even basic facilities for astronomy by some of the most influential American individuals and organizations of the day speak to the dearth of interest and support for astronomy prior to Adams's presidency. Within a few years of Adams's "light-houses of the sky" inaugural address and national observatory proposal, the first large aperture telescope would be imported from Europe to Yale. The first cluster of American observatory construction in the late 1830s coincided with the renewal of Adams's campaign for astronomy in connection with the Smithson bequest. Most tellingly, Adams played a key role in three of the four large observatory projects of the 1840s: his efforts were a major, if indirect, impetus to the establishment of the U.S. Naval Observatory; he was directly involved in fund-raising for astronomy at Harvard; and he was honored and feted at the inauguration of the Cincinnati Observatory, for engendering the new public esteem of astronomy that had allowed for the observatory's establishment. Adams's contribution to the development of American interest in the exploration of space was extensive. But there were also other forces that propelled American interest in astronomy in the mid-nineteenth century, forces that were arguably more basic and widespread among the population of America than either the lofty motives that buttressed Adams's support for astronomy or the scientific interests of professional astronomers, and it is to an exemplar of these that we now turn.

From the mid-1830s, and for the next four decades, the construction of observatories accelerated rapidly as part of what has been referred to as the "American Observatory Movement."[47] Although the general chronology and personal narratives of the leading individuals of the movement have been reasonably well sketched, there has been comparatively little attention paid to the question of how the necessary financial support was mobilized to give this movement its momentum. The focus here is to explore that question through examining the dynamics behind the founding of a number of individual observatories. As a backdrop, however, it is also valuable to look briefly at how the exploration of the heavens was perceived within the mass culture of the period. Such an investigation could fill volumes, and the subject is indeed deserving of such a comprehensive treatment. Although the examination here is only one example of how large observatories were perceived in the popular culture of the time, it is a particularly revealing one: the "Great Moon Hoax" of 1835. This is not a representative example by any means. On the contrary: it was a unique and era-marking event, the circumstances of which throw into high relief some of the contemporary trends in the popular perception of astronomy.

New York City in the 1830s was the urban center of commercial growth for the young American republic. It had a population of two hundred thousand to three hundred thousand, and this mass of people and commerce had created a new type of popular newspaper, the penny paper, which appealed more directly to the broader public than the six-cent, business-focused dailies. The editor of one such New York penny paper, *The Sun*, Richard Adams Locke, was an avid reader of natural science and astronomical literature. His readings included the popular works of Reverend Thomas Dick, a Scottish minister who appealed to religious sentiment and proclaimed that the myriad planets of the universe, including the Moon, had been purposely populated by God. Locke also read more scholarly texts, such as John Herschel's *A Treatise on Astronomy*, first published in America in 1834.[48] These readings led Locke to think about how a discovery of extraterrestrial life on the Moon would be received by the public, as well as the effect that such an announcement would have on newspaper sales.

On Tuesday, August 25, 1835, *The Sun* ran a front-page story announcing "GREAT ASTRONOMICAL DISCOVERIES Lately Made By Sir John Herschel, LL.D., F.R.S., &c, At the Cape of Good Hope." In it, *The Sun* claimed that it had been furnished with a recent issue of *The Edinburgh Journal of Science* that contained an article on Herschel's latest, amazing discoveries, which it was now reprinting. The story progressed over six installments for a week, culminating in the revelation that a powerful new telescope of Herschel's, with a magnifying power forty-two thousand times that of an unaided human eye, had enabled him to discover an inhabited world in the Moon, complete with mountains of amethyst, bipedal beavers, and a lunar civilization of intelligent, temple-building beings with batlike wings:

> Certainly they *were* like human beings, for their wings had now disappeared, and their attitude in walking was both erect and dig-

nified. Having observed them at this distance for some minutes, we introduced lens H z which brought them to the apparent proximity of eighty yards. . . . About half of the first party had passed beyond our canvass; but of all the others we had a perfect distinct and deliberate view. They averaged four feet in height, were covered, except on the face, with short and glossy copper-colored hair, and had wings composed of a thin membrane, without hair, lying snugly upon their backs, from the top of their shoulders to the calves of their legs. The face, which was of a yellowish flesh color, was a slight improvement upon that of the large orang outang, being more open and intelligent in its expression, and having a much greater expansion of forehead.[49]

Although the story was fictional, the effect it had was not. The account seems to have been largely believed to have been true when first published, and the widespread interest instantly increased the circulation of *The Sun* fivefold, giving it the largest circulation of any New York newspaper of its day.[50] The demand for news of the "discoveries" was staggering: P. T. Barnum, in his *The Humbugs of the World,* published in 1866, estimated that over $25,000 worth of "Moon Hoax" materials—which included the penny newspapers, sixty thousand twelve-cent pamphlets, and lithographs costing a quarter—had been sold by *The Sun,* more than had yet been spent on any American observatory.[51] There was more than fleeting interest: in 1859, a single copy of the twelve-cent pamphlet form of the story sold at a library auction for $3.75—roughly three or four days' average wage at the time.[52] The story made it over to Europe, where it was reprinted, sometimes with newly commissioned lithographs, causing correspondents to write to Herschel in South Africa seeking to affirm its validity. Herschel would later write of being told that a priest had even sermonized to his flock that he wanted to take up a collection for Bibles that could be sent to the newly discovered lunar inhabitants.[53] Although the tale of a "Bibles to the Moon" campaign in the 1830s may be apocryphal, it is suggestive of the significant impact, credibility, and influence of the manufactured story.

It is worth noting that contemporaries viewed the Moon Hoax as emerging out of a general popular interest in astronomy that was already present in America, rather than as being a major spur to such interest afterward. In P. T. Barnum's colorful and underutilized account of the hoax, it is evident that popular American interest in the exploration of the heavens predated the hoax and enabled it to reach the proportions that it did. As Barnum recounts:

> The real discoveries of the younger Herschel, whose fame seemed destined to eclipse that of the elder sage of the same name, and the eloquent startling works of Dr. Dick . . . did much to increase and keep up this peculiar mania of the time, until the whole community at last were literally occupied with little else than "stargazing." Dick's works on "The Sidereal Heavens," "Celestial Scenery," "The improvement of Society," etc., were read with the utmost avidity by rich and poor, old and young, in season and out of season. They were quoted in the parlor, at the table, on the promenade, at church, and even in the bedroom, until it absolutely seemed as though the whole community had "Dick" upon the brain. To the highly educated and imaginative portion of our good Gothamite population, the Doctor's glowing periods, full of the grandest speculations as to the starry worlds around us, their wondrous magnificence and ever-varying aspects of beauty and happiness were inexpressibly fascinating. The author's well-reasoned conjectures as to the majesty and beauty of their landscapes, the fertility and diversity of their soil, and the exalted intelligence and comeliness of their inhabitants, found hosts of believers. . . . It was at the very height of the furor above mentioned, that one morning the readers of the "sun" . . . were thrilled with the announcement in its columns of certain "Great Astronomical Discoveries."[54]

Although the words of a promoter and marketer like P. T. Barnum must be read with a healthy dose of skepticism, he was also a connoisseur of fads and chicanery and, as such, would have paid close attention to such a grand "humbug" as the Moon Hoax. That he confidently ascribes its success, and the credulity toward lunar habitation, to an already existing interest in astronomy is significant.

In referring to this general interest, in particular, as generated by the speculative works of Reverend Dick, it is hard to improve on the words of the contemporary commentator William Griggs: "So thoroughly was the popular mind, even among the best educated and most reading classes, imbued with these fanciful anticipations of vast lunar discoveries, that, at the time Mr. Locke's 'Moon Story' was written, scarcely any thing could have been devised and announced upon the subject too extravagant for general credulity to receive."[55] This ferment in popular astronomy was evidently significant enough that at least two of the era's prominent writers tried to capitalize on it with works of satirical fiction: Locke's "Moon Story" was the most successful, but Edgar Allan Poe had hoped his story "The Unparalleled Adventures of One Hans Pfall" might have similar success. Indeed, Poe initially accused Locke of plagiarism. Poe claimed to have considered writing a similar hoax-tale of telescopic discovery before choosing instead to develop his "Hans Pfall" story as a verisimilitudinous voyage to the Moon and publish it in a literary journal, the *Southern Literary Messenger*, in June 1835, a few months before *The Sun* ran Locke's work.[56] For his part, Locke claimed not to have read "Hans Pfall," and, after meeting him in 1843, Poe believed he was telling the truth. That two authors had apparently simultaneously generated their plans for using the popular enthusiasm for astronomy to their advantage suggests something of the depth and prevalence of that enthusiasm in American communities from New York to Baltimore.

An important and underappreciated factor in the success of the Moon Hoax was the description Locke provided of the scale of Herschel's telescope. Unlike in "Hans Pfall," which portrayed a voyage to the Moon being accomplished with a device of only moderate expense, Locke's story involved a new type of massive telescope that had been constructed at unprecedented cost. Locke sought verisimilitude for his story, and he achieved it through a superficially plausible method for a high-magnification telescope: combining the magnification techniques of the recently introduced hydro-oxygen microscope with a large Herschelian telescope. The stupendously large magnifying power was naturally accompanied by an impressively high cost: £70,000, an amount supposedly raised by subscription. The subscription effort was said to have been initially launched with

£10,000 from "that liberal patron of science" the president of the Royal Society, Prince Augustus Frederick, the Duke of Sussex, and ultimately backed by the king up to the full amount.[57] To the American reader, £70,000, converted to U.S. dollars at the then-official exchange rate of $4.85 to the pound, would be equivalent to some $340,000. Such an amount represents a share of contemporary wealth that would not be seen in terms of American observatory expenditures until some forty years later with the Lick Observatory. The supposed physical scale of the telescope was also unprecedented: an object glass 24 feet in diameter, six times the size of William Herschel's largest, suspended between pillars 150 feet high. That claims of a project of such unprecedented scale in astronomy would lend credibility to a narrative of such sensation and novelty is obvious. What is perhaps more interesting to note is that an astronomical project of such scale was itself deemed credible to a broad cross section of the population of New York in 1835.

Locke's fictional narrative also included construction details that were familiar themes for large observatories and that would have imparted credibility to the story. The casting of the large object glass was given due attention, with the work being done by "the large glass-house of Messrs. Hartly and Grant" and with the first glass found to have been significantly flawed, requiring another casting. The transportation of the equipment to Herschel's observatory in South Africa is described in some detail, with thirty-six oxen and several companies of Dutch Boers taking four days to traverse the distance from the port at Cape Town to the observing site. The telescope within the observatory was moved on circular railroads by a "locomotive apparatus" to enable maximum precision.[58] Locke even supplies the project's patron, the king, with the politically appropriate concern for the practical benefits of the project: "His Majesty, on being informed that the estimated expense was £70,000, naively inquired if the costly instrument would conduce to any improvement in *navigation?* On being inform that it undoubtedly would, the sailor King promised a *carte blanch* for the amount which might be required."[59] The story of the Great Moon Hoax shows the popular interest in astronomy and telescopic exploration that existed in America at the dawn of the Observatory Movement. Although the telescopes that would follow in the subsequent decade would be initially modest, the enthusiasm of New Yorkers for telescopic discoveries presaged the fervor for astronomy that would later grip the citizens of Cincinnati, Boston, Albany, Detroit, Pittsburgh, and Chicago and was an early sign of American fascination with large telescopes. By the end of the century, Americans would be building real telescopes that would almost rival Locke's fictional one in size, expense, and social impact.

Before his journey to Cincinnati, in an address to his constituents in Dedham, Massachusetts, Adams summed up his understanding of the economics of astronomy and the exploration of the heavens:

> Other sciences may be cultivated by individual exertion, by solitary toil, and at little cost—but for the discovery and investigation of the secrets of the skies, expensive edifices, still more expensive and complicated instruments, the combined labors of exquisitely talented mechanics, of eagle-eyed observers, of profound and skilful mathematicians, are all indispensably necessary; and, without the fostering aid and encouragement of the powerful, the affluent, and the liberal, these cannot be obtained. The history of astronomy has been, in all ages, the history of Genius and Industry, in their blazing light and untiring toil, patronized by power.[60]

It would indeed be the powerful and the affluent that would patronize astronomy in nineteenth-century America. But it would not be through the federal government, as Adams had initially expected, that American citizens would provide their support. Rather, it would be through the voluntary and personal decisions of individual Americans in pursuit of their own passions and their own monuments.

The early private support of American astronomy was focused in the colleges. In 1828, Yale College received a donation of $1,200 from Sheldon Clark, a bachelor who had been deprived of a much-desired education but who had been left significant property holdings by his grandfather. The gift enabled purchase of the finest ten-foot-long, five-inch-aperture Dollond achromatic telescope from London. Although it was not provided its own observatory and was placed instead in the steeple of the college's first chapel, the Atheneum, it was the finest telescope in the Americas. It soon established the importance of large telescopes on the continent with its 1835 sighting of Halley's Comet weeks before news arrived of its sighting in Europe. Popular interest in astronomical observatories was gathering momentum, and by the 1840s and 1850s, colleges would be acquiring telescopes as a way of signaling their intellectual bona fides, with even the Central Masonic Institute of Selma, Alabama, having a telescope as large as Yale's.[61]

Contemporaneous with the Yale telescope acquisition, Joseph Caldwell, the president of the University of North Carolina, initiated the construction of an observatory that would be the most expensive to date in the United States. As well as being the president of the university, Caldwell had a passion for teaching astronomy, as an 1896 article in *Popular Science* recounted: "To study the constellations and to show them to his pupils, Dr. Caldwell built on the top of his own residence a platform surrounded by a railing. Here he would sit night after night, pointing out to the seniors, taken in squads of three or four, the outlines of the constellations and their principal stars, and the highway of the planets and the moon."[62] He recommended that the University establish a fully equipped observatory and was passionate enough about the venture to make a trip to London to procure the instruments at his own expense. The trustees arranged a credit of $6,000 in 1831, which allowed Caldwell to purchase the finest Troughton and Simms meridian circle, an altazimuth telescope, a Dollond equatorial refractor, and a Molyneux clock with a mercury-compensating pendulum.[63] He built the observatory building with his own personal funds, for $430.29, although he was later reimbursed by the university. The observatory fell into decay shortly after Caldwell's death in 1835, as his successors did not see the value of such a costly astronomical investment being maintained at a university whose main aim was to train the would-be political and commercial leaders of the South.[64] However, Caldwell's efforts are a significant early example of the extent to which individual initiative and private passion can propel the development of astronomical observatories, if not necessarily maintain that momentum.

Another university president who gave significant support to astronomy was Willbur Fisk, the first president of Wesleyan University in Connecticut. A course in astronomy was available from the time of the university's founding in 1831, and considerations for a college observatory began as early as the following year. Fisk was a renowned Methodist minister and theologian, dedicated to education, who went to special lengths to incorporate a study of the heavens into his university's curriculum. In 1835 he traveled to Europe to order a six-inch Lerebours achromatic telescope in Paris for 6,000 francs, and the following year he purchased, for $2,200, a giant orrery weighing one ton, with a forty-five-foot-diameter orbit for Neptune and five hundred cog wheels of brass.[65] Total expenditure on "the philosophical and astronomical apparatus" in 1836 was estimated at $4,000, including a separate altazimuth telescope and astronomical clock.[66]

The Hopkins Observatory, at Williams College in Massachusetts, the oldest observatory still existing in the United States, was also largely the result of a personal passion for astronomy, in this case that of Professor Albert Hopkins, the brother of the university's president. In 1836, with $4,000 from the college, $500 of Hopkins's personal funds, $1,200 from the college trustees, and $400 from town merchants—a total of $6,100—a dedicated observatory was constructed to house an imported equatorial Herschelian telescope. As was common in many early American observatories, religious motivation was a significant factor. For Albert Hopkins, an observatory was a way to explore the heavens as the work of God, as his dedicatory oration made clear: "It is the desire of those whose contributions and whose care have aided in the erection of this building, that it may subserve the interest, not merely of sound science, but of spiritual religion."[67] Indeed, religious study and thought formed the greater part of Professor Hopkins's life, with astronomy being one part of that study. The religious motivations for the observatory were evident even in its architecture, with biblical phrases engraved into marble tablets above its doors. Over the north door was written, "Lift up your eyes on high and behold who hath created these."[68]

Although intrinsic interests had planted the seeds of the Observatory Movement, signaling motives began to assume importance even in the early phases. The Hudson Observatory at the Western Reserve College in Hudson, Ohio, was authorized in 1836 with $4,000, only ten years after the college itself had been founded with $7,500.[69] The funds covered a year of travel to Europe for Elias Loomis—the mathematics professor leading the project, who had learned astronomy at Yale with the Sheldon

Clark telescope. His European trip was for the purpose of acquiring the instruments for $1,700, and $1,086 more was set aside for the construction of the observatory.[70] For Loomis, the motivation was astronomical research, while the college saw an observatory as a requisite element of signaling their position as a prestigious educational institute. As explained by the later college president, Reverend Carroll Cutler, this signaling motive reflected a desire to emulate the astronomical prowess of other great American colleges such as Yale: "If the question were asked whether these buildings were all necessary, we should have to reply that the plan on which Yale College was conducted was adopted here as the sum of all wisdom in such matters."[71] The positive signaling that applied to buildings like observatories seemingly did not extend to the quality of professors, however, as Loomis was subject to "pecuniary embarrassments" at the university and decided to resign from the college in 1844 to accept a position at New York University.[72] The top quality of Loomis's research and teaching abilities can hardly be questioned: the textbooks he would later write would become so popular at universities and with the public that, at his death in 1889, he was able to leave a significant fortune, $300,000, to his alma mater, Yale University, for the exclusive support of its astronomical research program.[73] Already then, there were signs that the popular demand for astronomical observatories was driven by something other than a desire to support scientific research.

The Philadelphia High School Observatory, while pedagogically focused, foreshadowed the civic observatories to come and the role of signaling motivations in their establishment. In 1836, the federal government redistributed to the states some of the surplus from the national budget. The return of Halley's Comet in 1835 had increased general interest in the study of the heavens. This provided fertile ground for members of the American Philosophical Society and for George M. Justice, a prominent Quaker merchant who was a member of the Board of School Controllers, to cultivate the idea of using the funds for a research observatory to be established at Philadelphia's Central High School.[74] Like most American observatories of the day, the High School Observatory project selected a large equatorial refractor, ideal for exploring and examining the heavens rather than just measuring them. With members of the American Philosophical Society acting as technical advisors for the project, the high school became the first to import a German telescope, a seven-inch refractor from Merz and Mahler costing $2,200. With total funding of $5,000 allocated for the observatory, it was for a time the best equipped in the country.[75] Although the equipment was unrivaled in the United States, support for scientists to use that equipment for research was limited. The principal instructor in astronomy at the high school, Professor Ezra Otis Kendall, was unable to convince the controllers to allow him to shed some of his teaching duties to focus on research.[76] In contrast, support had been forthcoming when it was proposed that the instruments be mounted on marble pillars. The objective of the project, with the large observatory dome situated directly atop Central High School, was to make a powerful statement. This was not only an educational endeavor but also a monument to the civic spirit of Philadelphia, an organizing motive that would be prominent in later observatories in Cincinnati, Boston, Albany, Detroit, Pittsburgh, and Chicago.

As well as being an interesting early example of civic support for astronomy, the Philadelphia High School Observatory is a little appreciated milestone in the development of the American Observatory Movement. Not only was it the first observatory to have been planned with input from a wide range of members of the American scientific community, it also directly influenced the development of subsequent observatories. Based on his experience as controller for the project, George Justice would be asked to help with the planning of two more observatories. The Central High School project also set the trend for German refractors, which were purchased for most of the major observatories over the next five years, including the U.S. Naval, Cincinnati, and Harvard College Observatories.[77] Of perhaps even greater import, however, would be the influence that alumni of the school would have on the development of two of America's most important astronomical institutions. One alumnus, George Davidson, who taught as an assistant at the school's observatory, would be intimately involved with James Lick's decision to build his monumental observatory, while Charles Yerkes, who had learned astronomy from Professor Kendall, would go on to make his fortune and endow his own Yerkes Observatory with the world's largest refractor.[78]

An observatory located on top of a Philadelphia high school may seem an unlikely inflexion point in the history of American space exploration.

Yet its establishment was the culmination of an interest in and support for astronomy in Philadelphia that stretched back to pre-Revolutionary days and the 1769 transit of Venus effort, and it was the first significant observatory built in a major American city. The Philadelphia High School Observatory directly influenced the development of subsequent observatories, and it planted, through its alumni, seeds that would contribute to the growth of some of the most important "big science" projects of late-nineteenth-century space exploration.

The early observatories, from Yale to Philadelphia, were smaller in scale than those that would follow in the decades to come, but they established a number of the trends that would continue to be seen through the course of the later Observatory Movement: the importance of leading individuals in the establishment of observatories, the preeminence of private funding, the role of signaling and signal emulation, the growth of intrinsic interest in the exploration of heavens, the religious undercurrent behind much of the support for astronomy, the sense of civic identity and support that observatories could engender, and even the conflict between funding the scientific research that astronomers favored versus financing the spectacle of the observatory favored by the patrons.

Though universities would continue to be a focal point of American astronomy, the scale of the largest observatories would increasingly rely on private and civic motives that superseded academic and pedagogic ones. The Harvard College Observatory would be just as much the result of a desire on the part of the citizens of Boston to have a world-class observatory for the purpose of signaling their beneficence and wealth as it would be the result of the efforts of Harvard astronomers. Likewise, while the Lick Observatory was to be managed by the University of California, its impetus came from the desires of an individual who was fulfilling significantly broader ambitions than school patronage. While moderate-size university observatories, similar in motive and scale to the early observatories of the 1820s and 1830s, would continue to be a mainstay of American observatory construction, the largest and most expensive telescopes became increasingly intertwined with broader national, regional, dynastic, and even international forces. Within the first five years of the 1840s, four major observatories would be founded, each at a scale and expense eclipsing the earlier college observatories, and each driven more overtly by signaling motivations.

The Astronomical Observatory at Georgetown College in Washington provides another interesting transition point from the college observatories, with their principally internal focus, to the civic observatories, where external signaling concerns and the implications for community identity became more overt organizing motivations. In the case of the Georgetown Observatory, however, the community was not one of civic-minded urban citizens, but rather a religious order that had long used astronomy as a signaling device. Although the case of the Georgetown Observatory has unique features, it was part of a larger trend of religious sentiment providing significant support for astronomy. Astronomy was an integral part of the natural theology of the period, with the immensity and order of the universe, as revealed by astronomy, being widely interpreted as a sign of God's handiwork. The intrinsic motivations of religious belief thus played a significant role in the funding of early American observatories, and the intertwining of religion and astronomy can be seen in the writings of a number of influential early Americans.

One of the earliest expressions of this linkage comes from Benjamin Franklin, who was a committed Christian with religious views that were shaped by the discoveries of astronomy. In the opening lines of his "Articles of Belief and Acts of Religion," written in 1728, Franklin reveals his fascination with the cosmos: "When I stretch my Imagination through and beyond our system of planets, beyond the visible fixed stars themselves, into that space that is every way infinite, and conceive it filled with suns like ours, each with a chorus of worlds for ever moving round him, then this little ball on which we move, seems, even in my narrow imagination, to be almost nothing and my self less than nothing, and of no sort of consequence."[79] For many Americans in the nineteenth century, the wonders of the cosmos were seen as obvious proof of the existence of a deity. For support of this worldview, American intellectuals could point to the man then considered the greatest natural philosopher of all time, Sir Isaac Newton: "This most beautiful system of the sun, planets, and comets, could only proceed from the counsel and dominion of an intelligent and powerful Being. And if the fixed stars are the centres of other like systems, these, being formed by the like wise counsel, must be all subject to the dominion of One. . . . This being governs all things, not as the soul of the world, but as Lord over all; and on account of his dominion he is wont to be called *Lord God* or *Universal Ruler*."[80]

Seeing God in the universe would have seemed as obvious to a nineteenth-century American as the changing seasons. Through this perspective, astronomy was often seen as a humbling reminder to humanity of its status within a grand plan. David Rittenhouse described this sentiment eloquently: "All yonder stars innumerable, with their dependencies, may perhaps compose but the leaf of a flower in the creator's garden, or a single pillar in the immense building of the divine architect."[81] Such thoughts and writings form a thread in American intellectual life, which can be traced from the Puritans through the time of Franklin to the nineteenth century, creating a strong natural alliance between astronomy and religious sentiment—one that could be leveraged to attract resources to astronomical endeavor.

As we have already seen, John Quincy Adams was one such American who saw stargazing as an activity directly analogous to worship. He firmly believed that "the heavens declare the glory of God; and the firmament sheweth his handywork" (Psalm 19:1) and that an observatory was therefore a "Temple hallowed to the worship of the Creator, raising the Souls of all who are admitted to its nightly disclosures, to a more intimate communion with the author of the Universe and with the ever multiplying wonders of his creation."[82] With sentiments like these permeating the thoughts of American intellectual leaders, it should not be surprising that members of one of the most powerful religious orders, the Society of Jesus, would look to build such a temple in America.

The tradition of Jesuit observatories is an example of the strong linkage between astronomy and religion, as well as the use of observatories as signaling devices. At their height in the nineteenth century, there were fifteen Jesuit observatories across North America, with the most important being in Georgetown, St. Louis, Boston, and Montreal.[83] The Jesuit order had long made astronomy a key part of their education system, as well as part of their proselytizing activities. The signaling value of astronomical observatories was explicitly understood and utilized by the Jesuits.

As Agustin Udias notes, "the scientific prestige of the observatories was considered to be an important factor in spreading the Christian message."[84] In the nineteenth and early twentieth centuries, the Jesuits had seventy-four observatories in missions as far afield as Paraguay, Madagascar, China, Zimbabwe, the Philippines, Lebanon, Colombia, and Australia.[85] Although many of these efforts were relatively modest in scale, a number constituted significant expenditures. For instance, a new building for the Manila Observatory was erected in 1894, complete with the latest imported instruments, at a cost of $40,000.[86] Although these observatories were often the results of efforts by dedicated individuals committed to scientific investigation and personal exploration of the heavens, these structures, and the insights into the cosmos that they enabled, were also used by the Jesuits to signal to the world the strength, sophistication, and high-mindedness of the order.

The Georgetown Observatory was the first, largest, and most important example of this Jesuit tradition in the United States. Like with earlier university observatories, the faculty initiated its establishment, but, as with later observatories, it would quickly acquire a significance extending beyond its campus—in this case, all the way to the Vatican.

In 1841, an Irish Jesuit, Father James Curley, made his case for an observatory at Georgetown and solicited donations from other American Jesuits, eventually persuading the young Thomas Jenkins to offer $8,000 from his inheritance. Reverend Charles Stonestreet offered another $2,000 the following year, and the Jenkins family would make additional contributions. Curran and O'Donovan estimate the building cost at $9,000 and the instruments at perhaps twice that amount, for a total of $27,000, making it one of the most expensive observatories of the era.[87] Yet, despite the funds coming almost exclusively from American Jesuits and their families, there was opposition to the project from Rome. The procurator of Georgetown thought it "a true folly" but felt it "worse than useless" to preach fiscal prudence amid such enthusiasm.[88] In Rome, the superior general of the Jesuits, Jan Roothaan, was similarly concerned and wrote to Curley denying approval for the project. Curley, however, through some curious misinterpretation of Roothaan's response, continued with the project. When he was ordered to defend the project again to Roothaan, he explained that it was "building public esteem for the college's commitment to science" and thus attracting additional contributions to other causes supported by the order.[89] The superior general, however, remained concerned that such a monument might be seen as an extravagance and again instructed Curley to halt the project. The letters going back and forth to Rome would take weeks to travel each way, however, and the observatory was virtually completed by the time Roothaan's final words on the matter reached Georgetown in 1844.

Although the Jesuits in Georgetown and the Jesuits in Rome had opposite views of the matter, they were both concerned with the signaling value of the observatory. Reverend Curley thought that the observatory would signal the strength of Georgetown and of the Jesuit order in America, while Superior General Roothaan was specifically concerned with what message such an expensive signal sent. Once established, however, Rome supported the observatory, sending its top astronomers, Father Francesco de Vico, director of the Vatican Observatory, and Father Angelo Secchi, a pioneer of spectroscopy, to the observatory during the revolutions in the Italian states in 1848. The arrival of two notable European astronomers was itself interpreted as a signal of growing American intellectual prestige and was heralded by the American press as a coup for the nation. Not since the transit of Venus observations in 1769 had American science garnered such international attention, and never before had the signaling ambitions of an international organization been interwoven with the founding of an American observatory. Although it may have been religious devotion and a personal desire to explore the heavens that had initiated the Georgetown Observatory project, the cost and visibility of the observatory turned it into a project of prestige for the order.

The Georgetown Observatory project demonstrates how the motivations of signaling, science, and religious devotion could combine to spur forward the construction of astronomical observatories. Similarly, the observatory attached to the Friends' Central School in Philadelphia became a point of pride for the Philadelphia Quaker community and was appreciated as a visible signal of their high-mindedness.[90] Built in 1846 with a fifteen-foot revolving dome and a five-inch equatorial refractor, it was a nontrivial expenditure for its patron and director, Miers Fisher Longstreth. While it did serve as a signal for the Quaker community, Longstreth's motivation was a personal desire to study the work of God in the heavens.

This personal desire to study the natural theology of the heavens was the original motivating force for many of the projects that marked the American Observatory Movement, not only for religious societies but in the founding of personal, university, and civic observatories as well. The story of the Georgetown and the Friends' Central School Observatories also elucidates a more general trend: the process by which projects are initiated out of the personal interests of individuals and transformed into signaling devices for broader communities. This process would be repeated many times throughout the history of American space exploration, including in the founding of the nation's first national observatory.

The founding of the U.S. Naval Observatory in the early 1840s presents another example of continuum in the history of U.S. space exploration and a pattern that will be seen even more emphatically in later chapters— that of individuals leveraging and shaping military demands in order to secure funding for facilities and projects that are their personal objectives. A. Hunter Dupree famously called the Naval Observatory "the classic example of the surreptitious creation of a scientific institution." There is much to support this statement in the way in which a few individuals engendered the growth of the Naval Observatory, rather unexpectedly, from a simple, privately funded storage and rating depot for marine chronometers to an institution hailed as the nation's first "National Observatory." However, as Steven J. Dick rightly cautions in his authoritative history of the institution, the transformation from depot to national observatory was also the result of broader forces running through the politics, culture, and military bureaucracy of the time, including John Quincy Adams's observatory proposal and the seemingly irresistible pull of a national observatory as a mark of prestige.[91]

In the nineteenth century, astronomy was a crucial skill for surveying and navigation, and thus it received modest federal patronage from a variety of sources, one of the most prominent being the U.S. military. Early support came from the necessity of training young officers in these requisite skills at West Point Academy. Three towers for astronomical observation were part of the academy's large library building, which initially housed a 6-inch refractor until it was replaced in 1842 by a $5,000 American Henry Fitz refractor of 9.75 inches.[92] The military institution that would become America's "national observatory," however, started out even less auspiciously as a small personal initiative.

The U.S. Navy's Depot of Charts and Instruments was established and expanded through the personal ambitions of three young officers. In 1830, Lieutenant Louis M. Goldborough proposed a depot to store marine chronometers and provide the critical service of rating their accuracy. This involved calculating the error in timekeeping for each chronometer in comparison to the time as measured by the movement of the sidereal heavens, so that the error of each chronometer could be factored into navigators' calculations to determine their longitude at sea. The business of rating chronometers thus involved astronomical instruments, principally a high-quality transit instrument for timing sidereal movement. With the help of his father, then the secretary of the Board of Navy Commissioners, Lieutenant Goldborough managed to secure an initial annual budget of $330 for a small depot facility of which he was to be the sole employee.[93] Goldborough's successor, Lieutenant Charles Wilkes, moved the depot to a new location on his personal property in 1833 and constructed a small observatory there for observations.

It was the personal ambition of the depot's third head, Lieutenant James Melville Gilliss, that would increase the scope of the depot to include astronomical observation beyond that which was required by the navy's operational needs. Gilliss saw expanding the depot's activities to include astronomical research in general as a way to improve the technical reputation of the Navy: "I should have regarded it as time misspent to labor so earnestly only to establish a *depot*. My aim was higher. It was to place an institution under the management of *naval officers,* where, in the practical pursuit of the highest known branch of science, they would compel an acknowledgement of abilities hitherto withheld from the service."[94] Motivated to improve the reputation of the navy, Gilliss initiated a major proposal for a new depot that would include a permanent astronomical observatory.

Although Gilliss himself was interested in developing astronomy within the navy to signal the capabilities of its officers, his proposal was couched in explicitly utilitarian terms. His appeal to the board in 1841 rested on a number of factors: the experience and utility of the depot's chronometer rating, the unsuitability of the existing building for the task, the uncomfortable fact that the property now occupied by the depot was privately owned, and the defects of the current transit instrument. No stress was laid on the facilities for general astronomical observation that had been included in the request. The board approved the proposal and forwarded it to Congress, and there it languished. Congressional inaction may have been due to lingering opposition to the Adams proposal, which had been recently reborn in the debate on the Smithson bequest. Interestingly, it was the opposition to Adams's proposal, however, that became an unlikely source of support for an observatory-equipped depot, specifically because the chair of the Senate Naval Committee, Senator W. C. Preston, recognized and disapproved of the signal that a genuine national observatory would send.

Senator Preston had originally opposed even accepting the Smithson bequest, believing that it would increase federal power at the expense of the states, and, not surprisingly, he adamantly opposed Adams's idea of using the funds for an observatory, which would be seen to signal federal power. Preston's position took a tactical shift, however, after Gilliss made a presentation to the National Institute for the Promotion of Science, which Senator Preston favored as a potential destination for the Smithson bequest. When Gilliss advocated the idea of expanding the Naval Depot of Charts and Instruments by adding a small observatory, Preston calculated that an observatory within the navy would undermine Adams's case for using the Smithson bequest to fund a larger national observatory and thereby strengthen the case for his proposed use of the funds.[95] With Preston's support, the Senate passed a bill endorsing the depot proposal in 1842, and the House soon followed suit, authorizing an impressive $25,000 and appropriating $10,000, leaving the remainder of the funds and the details to be provided by the navy.[96] With the depot's observatory signed into law, Gilliss and the navy were free to use the ample authorization to outfit an observatory with the finest astronomical instruments and an imposing edifice. Even after a shuffle in naval leadership left the hydrographer Matthew Fontaine Maury in charge of the newly completed depot in 1844, the value of the observatory as a route to naval prestige was obvious. Maury, with the assistance of the secretary of the navy, George Bancroft, seized the title of "National Observatory" for the new facility simply by declaring it so on the title page of the depot's first published volume of astronomical observations.[97]

Although it was never officially designated the "National Observatory," it became widely recognized as a de facto one, perhaps most significantly by Adams. In the final debate on the Smithson bequest in 1846, Adams stated, "I am delighted that an astronomical observatory—not perhaps so great as it should have been—has been smuggled into the number of institutions of the country, under the mask of a small depot for charts," and thus concluded, "I no longer wish any portion of this fund to be applied to an astronomical observatory."[98] Through the efforts of Maury and Alexander Dallas Bache, the observatory would soon attract some of the best American astronomers and remained a leading American center for the exploration of the heavens until the early twentieth century. Throughout that time, the observatory's reputation as a national symbol would continue to be a dominant theme. The observatory's superintendent, Rear Admiral Benjamin F. Sands, supported the purchase of the world's largest refractor for the observatory in 1870, at a cost of $50,000, and lobbied hard for the even more expensive transit of Venus expeditions in 1874 and 1882, largely on the basis of national prestige and navy pride.[99]

And prestige and pride did flow. Elias Loomis would note that the astronomical work "placed our National Observatory in the first rank with the oldest and best institutions of the same kind in Europe."[100] James Ferguson's discovery of the asteroid 31 Euphrosyne at the observatory in 1854 became a point of particular pride, as would Asaph Hall's discovery of the two moons of Mars in 1877. The original motivation for Preston and Congress supporting the Naval Observatory may have been to put an end to Adams's campaign for a national observatory as a symbol of federal prestige and power. Cultural and institutional forces nonetheless ultimately shaped the Naval Observatory into the type of national symbol that Adams had desired. It has even remained a national symbol into the twenty-first century, as located on the observatory grounds—in the house built in 1893 for the observatory's superintendent—is the current official residence of the vice president of the United States.

That the U.S. Naval Observatory evolved toward its signaling role from its initially utilitarian one is an important indication of the signaling value ascribed to astronomical observatories in nineteenth-century America and of the financial support that could be mobilized on that basis. This equation—which we have already seen at play in the establishment of the early university and college observatories, in a religious-order manifestation with the Jesuits and Quakers, and in the development of the Naval/"National" Observatory—would yield even more impressive results when combined with the civic pride and boosterism of an emerging American power seeking to project itself on the world stage. What I refer to as the "civic observatory movement" would mark the apogee of popular nineteenth-century American enthusiasm for astronomical space exploration. It would also move the endeavor from the level of intellectual pursuit to that of mass culture, drawing on the inherent potential for doing so, which we have already seen in the Moon Hoax episode. The following chapter traces both this flow of funds into civic observatories and the parallel stream of what I refer to as "founder" observatories—those that were funded largely through the benefaction of a single patron. In time, it would be founder observatories that would come to dominate American astronomy. This would be due in part to the increasing wealth inequality in the Gilded Age, but also because their wealthy patrons, while also motivated by signaling interests, were interested in legacy—reputational signaling to future generations—a long-term focus that aligned well with the long-term research interests of professional astronomers. These two funding sources—community patronage and individual wealthy patrons—both generated largely in the private sector, and both driven primarily by signaling motives, combined to produce an impressive nineteenth-century U.S. space exploration effort. Quantifying its magnitude is our next task.

2

PUBLIC SPIRIT AND PATRONAGE: AMERICAN
OBSERVATORIES

It is also a place where men of business may acquire new ideas of the
wonders of the material universe; where men, whose days are spent in
toiling for the acquisition of wealth, may learn that there are miens of
intellectual riches more inexhaustible than the mines of California.

—*Elias Loomis, "Astronomical Observatories in the United States," 1856*

The prominence of American civic observatories—beginning with
the Cincinnati and Harvard College Observatories in the 1840s, and continuing with the similarly motivated observatories in Albany, Detroit,
Pittsburgh, and Chicago—is evidence of a broad-based, popular interest
in astronomy and in the use of large telescopes as vehicles for the personal exploration of space. While sometimes associated with colleges or
universities, they are differentiated by the fact that they were motivated
and funded by a civic enthusiasm going far beyond the academic sphere.
Although often an outgrowth of genuine interest in the exploration of
the heavens, they were also used to signal the coming-of-age of a city, as
well as to boost the prestige of the nation as a whole. As the Boston cor-
respondent of *The Athenaeum* put it in 1840, "One of the prominent subjects of discussion among our *savants* is the establishment of *Observatories*
of a character suitable to our standing as a civilized nation."[1] As Miller's
general study of scientific patronage in the period observes, "astronomy
was the queen of science, and its cultivation a sure sign of cultural accomplishment."[2] The civic observatories reflected the degree to which this

sentiment had permeated both across America and through the layers of the social strata.

The Cincinnati Observatory is a particularly fascinating example of broad-based support for space exploration in the mid-nineteenth century. In the 1840s, Cincinnati was America's sixth largest city and was undergoing an economic and demographic boom, with the population increasing from 46,000 in 1840 to 115,000 by the end of the decade.[3] It was in the midst of this boom that, in the spring of 1842, Ormsby Mac-Knight Mitchel—a charismatic professor of mathematics, philosophy, and astronomy at Cincinnati College and a former assistant professor of mathematics at West Point—gave a public lecture series entitled The Planetary and Stellar Worlds. Through his lecture series Mitchel described these worlds to the citizens of Cincinnati. He was aided by visually stunning lantern slides of galaxies and nebulae, produced from drawings made at the Dorpat Observatory, which then housed the largest refractor in the world in what is now modern-day Estonia and was then the Russian Empire. His lectures were wildly popular, and after he had finished them he was asked to repeat his last lecture at the city's largest meeting hall, the Methodist Episcopal Wesley Chapel. There, at the end of his lecture, in front of an audience of two thousand, he made an appeal to the people of Cincinnati for a telescope to surpass that of the Dorpat Observatory. He lamented that "while Russia with its hordes of barbarians boasted the finest observatory in the world, our own country with all its freedom and intelligence . . . had literally done nothing."[4] He declared that if government patronage was not to be found and if the wealthy were "too indolent and too indifferent," then ordinary people would have to take up the cause of science in America.

Mitchel came up with a funding model that presented the proposed observatory as a shareholder corporation and community asset in which all those who helped fund its construction could personally share in the exploration of space that it enabled. He divided the $7,500 he felt was required to purchase the necessary refractor into three hundred equal shares of $25 each to be payable when the entire amount was subscribed.[5] With the $25 came membership in the Cincinnati Astronomical Society, which planned to grant members the privilege of looking at the heavens through the best telescope in the world. Mitchel evidently expected this model to appeal most directly to the burgeoning middle class of the city: "I will go to the people, and by the anvil of the blacksmith, by the work bench of the carpenter, and thus onward to the rich parlor of the wealthy, I will plead the cause of science."[6] This approach was remarkably successful, with initial subscriptions obtained from sixty-seven different professions including: thirty-nine grocers; thirty-four landlords; thirty-three lawyers; six judges; seven paperhangers; five steamboat owners; three stable keepers; three stonemasons; three butchers; two lamp dealers; two plumbers; and one brick maker, among others.[7]

In addition to selling direct exploration of the heavens to the general public, Mitchel was also calling on prestige and civic patriotism to produce the funds. In the preamble to the constitution of the Cincinnati Astronomical Society, a direct connection was made between the private support of astronomy and the republican government of America: "Realizing the truth, that in our own country, and under a republican form of government, the people must hold, with respect, to all great scientific enterprises, that position of patrons, which in monarchial governments is held by Kings and Emperors."[8] Mitchel also combined this patriotic sentiment with a sense of civic pride in the context of international competition: "I am determined to show the autocrat of all the Russias that an obscure individual in this wilderness city in a republican country can raise here more money by voluntary gift in behalf of science than his majesty can raise in the same way throughout his whole domains."[9] As we have seen earlier with John Quincy Adams, the anti-Russian rhetoric that pervaded the space race had interesting parallels in nineteenth-century America and was also at that time placed in the context of competition between two political systems.

Against rather long odds, Mitchel succeeded in his aim of obtaining general public funding for the project. Walking up and down the streets of Cincinnati with his subscription book, he secured $9,437 for the telescope and approximately $6,500 for the observatory building.[10] A local philanthropist donated a tract of hilltop land on which the observatory could be built. With the resources in place, as would be the case in Albany and Pittsburgh, an independent company was incorporated that would be solely responsible for the completion of the project and its subsequent management. That the required resources were raised wholly through public subscription is indicative of the appeal held by an observatory and by the prospect of direct access to best-in-class space exploration technology.

Although the popular appeal of astronomy and the exploration of the heavens is clear, it is less clear that the exchange the subscribers entered into had much to do with science from their point of view. The broad interest in the personal exploration of the heavens was so significant that it effectively precluded Mitchel's attempts to do research with the observatory. When Mitchel tried to limit visiting hours to the observatory in order to conduct his research, this was strongly opposed by the subscribers who had underwritten the project and who demanded that the observatory remain open so that they could use it as intended.[11] Tellingly, the subscribers had chosen not to provide funds for research materials other than the telescopes or for the salary of astronomer.

In the now aging John Quincy Adams's mind, the project had little to do with science in the first place. Although he did agree to be the keynote speaker at the observatory's opening, he was annoyed at Mitchel's "braggart vanity which he passes off for scientific enthusiasm" and accused him of expending more effort on a sumptuous edifice, pageantry, and "gloss of showy representation" than on scientific research.[12] This attack on Mitchel was somewhat misplaced, as it was in fact the patrons of the observatory who had most clearly demonstrated these priorities. For Adams the Cincinnati Observatory's principal value was signaling, lauding the citizen subscribers in his oration for connecting the honor of America "with the constant and untiring exploration of the firmament of heaven."[13] Miller has claimed that "the Cincinnati Observatory was an important symbol of the place of science in American life."[14] In light of the backlash against Mitchel's attempt to do dedicated scientific research, however, it would be more accurate to say that the Cincinnati Observatory was an important symbol of the desire by Americans for personal involvement in the exploration of the heavens and the social status that accompanied it.

As a symbol of prestige and a signal of the new growth and wealth of a frontier town, the Cincinnati Observatory also acted as a spur to the men of affluence in old Boston. At a lecture in the Boston Odeon on the recently observed Great Comet of 1843, one of the leading American scientists and mathematicians, Benjamin Pierce, addressed a crowd of a thousand on the topic of astronomy and the need for a modern observatory in Boston. He made an appeal similar to that which Mitchel had made to the citizens of Cincinnati: to fund by public subscription a new large telescope for the observatory at Harvard College. Numerous previous attempts had been made, including by John Quincy Adams, to raise funds for an observatory at Harvard, all unsuccessful. Eventually the college itself provided $1,000 in 1839 for a basic observatory for teaching at Dana House.[15] This time, however, a large telescope for research was the goal, and heavier emphasis was placed on civic pride and prestige, noting that Yale, and even a Philadelphia high school, possessed better instruments than Harvard did and that an upstart Ohio river town had now managed to raise over $10,000 for a large telescope.

Stirred by the spirit of civic competition, the citizens of Boston determined to procure for their city and university the largest telescope in America. With local textile tycoon Abbott Lawrence as chairman of the subscription committee and an initial pledge of $5,000 from prominent landowner David Sears for the erection of the observatory tower, a total of $25,000 was raised.[16] Although the base of support was not as wide and varied as it had been in Cincinnati, the funds nonetheless came from over ninety-five sources, including $4,000 from businesses and societies.[17] The analysis done by Marc Rothenberg on the Harvard College Observatory's funding shows that the main donors were members of the elite, learned societies, and corporations—including insurance companies, which hoped better celestial navigation methods might reduce shipping losses.[18] The new observatory project experienced significant cost overruns, with an expenditure of around $50,000 from 1843 to 1846, and there were additional fund-raising campaigns over the next eight years, contributing an additional $14,000 on top of the original $25,000.[19] Out of the forty wealthiest Bostonians at the time, almost half of them (nineteen) were donors to the Harvard College Observatory during the decade. The most impressive donation was that following the suicide of young Edward Bromfield Phillips, son of one of Boston's wealthiest and oldest families. He bequeathed to the observatory a massive $100,000 endowment in 1848 in consequence of his friendship with George P. Bond, the son of the observatory astronomer.[20] The scale of the philanthropy and the extended period over which the observatory remained a focus for the elite of Boston are testaments to the prominent role that astronomy held within Boston's civic society.

The new Harvard College Observatory presented an unprecedented opportunity for American astronomers to take the lead in the exploration of the heavens. However, as at Cincinnati, the motivations that had led to the telescope's funding had consequences that, at least temporarily, impeded research. The generous funding of the new observatory allowed for the purchase of a fifteen-inch equatorial refractor from Mertz and Mahler, equal to that of the Pulkovo Observatory, allowing it to share the title of largest refractor in the world. As the first astronomical instrument of such distinction in the Americas, there was enormous potential for discovery, some of which was certainly realized, one example being the pioneering work of William Cranch Bond and John Adams Whipple in producing early daguerreotypes of the Moon and the stars. However, a number of factors limited the ability of Harvard astronomers to capitalize on the telescope's potential. For one, as in Cincinnati, no provisions had been made for salaried positions at the observatory, with the astronomers relying on, and having to seek, ad hoc university funding until the Phillips bequest in 1849. Secondly, the civic nature of the observatory's funding also led to expectations by the general public that the observatory would be available for public use. Efforts were made to accommodate this expectation, with the rather astonishing result that the citizens of Boston, for a time, had relatively open access to use the largest refracting telescope in the world for their personal explorations of space. However, as the provision of this service significantly reduced the availability of the telescope for scientific investigation, the astronomers protested, asserting their priority within the university environment, and access was ultimately curtailed.[21] These conflicts highlight the important role that both signaling and the desire for personal access to the exploration of space had in securing funding for observatories, along with the rather secondary interest of supporting scientific research.

The third most significant example of the civic observatory, and the most infamous case of conflict between the desires of an observatory's patrons and its astronomers, is the Dudley Observatory established in Albany. By the mid-nineteenth century, land speculation and the Erie Canal had made Albany a center of wealth and culture. City patricians had high hopes that the city might become the intellectual center of America and planned for the establishment of the University of Albany with a world-leading, hilltop observatory figuring prominently in the plan as a visible sign of the New York State capitol's emergence onto the world stage. Due to his fame as the founder of the Cincinnati Observatory and his subsequent popular lecture tours, the university organizers contacted Ormsby MacKnight Mitchel to help establish the new observatory. While Mitchel laid out a general plan and made the initial cost estimate of $25,000, it was the local elite who organized the funds. The principal organizers were James Armsby, a prominent fund-raiser for philanthropic causes in Albany, and Thomas Worth Olcott, a powerful banker whose Albany Mechanics and Farmers Bank had funded the political machinery of President Martin Van Buren, New York Governor Charles Marcy, and Senator Charles Dudley. With their connections, $25,000 was raised from about twenty of Albany's most influential citizens within two months of the campaign's commencement, kicked off by a founding donation of $13,000 from Mrs. Blandina Dudley to memorialize her husband.[22] With a grant of land for the observatory from General Henry Van Rensselaer, and a promise of $1,500 annual salary for Mitchel, the project had as generous a foundation as any of the era.

Although it had a strong foundation, it was one based on civic sentiment and prestige, not on science, and in many ways it would be the very strength of this foundation that would lead to its troubles. When obligations required Mitchel to return to Ohio, the trustees of the observatory, few of whom had even an amateur's interest in astronomy, confidently proceeded with what they considered the matter of highest priority—the construction of a suitably impressive observatory building. Only when this had been completed in 1854 did the trustees turn their full attention to the secondary consideration of acquiring astronomical instruments and astronomers. That the project, meant to become a leading institute of astronomy, could proceed comfortably for years almost wholly without scientific input highlights the secondary nature of science and the prominence of the signaling motive in the founding of the observatory. When a scientific council was finally convened in 1855, it was appropriately replete with some of the most prestigious names in American science: Alexander Bache, director of the U.S. Coast Survey; Joseph Henry of the Smithsonian Institution; Benjamin Gould, who had studied under Carl Friedrich Gauss and was the first American to receive a Ph.D. in astronomy; and Benjamin Pierce, a prominent Harvard scientist. These influential men of science, however, were less than enthusiastic about the way that signaling had trumped science in the development of the observatory.

Although chroniclers of the dispute between the trustees and the science council have placed emphasis on personal relations and differing notions of what constituted gentlemanly conduct, there were real differences of opinion on the shape and purpose of the observatory.[23] The science council failed to appreciate the importance of an elegant edifice to the donors and the increased base of support that would come with a large equatorial telescope that enabled the citizens of Albany to make their own use of the observatory. Instead, the scientists demanded architectural design changes that were aesthetically disappointing and they chose as their principal instrument a fixed heliometer for scientific measurements of the Sun. While a heliometer would enable them to conduct work in their preferred field of solar observations, it prohibited satisfying public stargazing. The stage for conflict had been set.

The trustees wanted their observatory to be a leading institute of astronomy and so were initially deferential to the scientists' requests. To do so, however, required significant increases in funding and the provision of an endowment for the salaries of the astronomers. The widow-benefactor Mrs. Dudley was able to come up with the additional $14,000 for the heliometer in 1856 as well as $50,000 of an $80,000 endowment raised for salaries and operations, for a total of $119,000. The remainder of the funds were raised from William Astor, the trustees, and sixty other prominent men and women of New York.[24] Despite this generosity, scientific council members Pierce and Gould were uncompromising in their pursuit of what they considered "pure science." They further antagonized the donors with the firing of an assistant who had named a small, newly discovered comet after one of the most important benefactors, Thomas Olcott. The personal conflict between the two sides eventually escalated to the point at which Gould and Pierce barricaded themselves in the observatory and declared squatter sovereignty and rights as legal guardians. The trustees responded by terminating all agreements, threatening legal action, and distributing fifteen thousand copies of a 173-page polemic against Gould's character and behavior.[25] The trustees ultimately battered down the door of the observatory and threw Gould out into the January snow.

At the heart of this debate was the same conflict that had been seen on a smaller scale in Cincinnati and Boston, that between scientific research and the aesthetic and exploratory desires of those who had provided the funds. The type of instrument was at the crux in Albany, with Gould demanding a purely scientific institution with an instrument for cutting-edge research and with Olcott desiring a general-purpose telescope and at least some public access to the observatory. The trustees raised this point explicitly in their statement on the conduct of Gould: "Citizens, when visiting the Observatory and grounds were, in repeated instances, treated with incivility. Sometimes admission was refused altogether; and, at other times, when they succeeded in obtaining admission, visitors were received with so little respect or courtesy that they felt themselves at liberty to complain of their treatment."[26] For his part, Gould, in his lengthy response to the statement of the trustees, attacked them for the priority that they gave to superficial matters: "The empty dazzle of temporary show was, in the wishes of the managing Trustees, paramount to any ideas of scientific usefulness or dignity."[27] Gould also lamented the openness that Olcott and the trustees prized: The Dudley Observatory "ought not to be a place where every observation is interrupted by curious visiters [*sic*], who suppose the establishment to be a sort of exhibition, or where the instruments are continually liable to serious disturbance by idle and meddlesome fingers."[28] This disagreement between the trustees and scientists on the objectives and requirements of the observatory led to a 300 percent cost overrun and a fundamentally crippled institution. The principal legacy of the Dudley Observatory is unfortunately that of being the first major American space exploration boondoggle. Nonetheless, the extensive debate between the trustees and the scientific council makes clear the motives that had driven such an expensive project in the first place: a desire for a prestigious signal and a general interest in the personal exploration of the heavens.

There were a number of other examples of astronomical observatories that were funded through public subscription by civic pride and popular interest in astronomy. In 1848, in Shelbyville, Kentucky, relatively close to the recently finished Cincinnati Observatory, an impressive observatory building was constructed at Shelby College and a large 7.5-inch Merz and Mahler equatorial telescope, costing $3,500, was ordered, with funds raised at least partly by public subscription.[29] When Henry Tappan became the president of the University of Michigan, he set the development of an astronomy program as one of his highest priorities and made a plea for observatory funding at his inauguration in 1852.[30] Twenty-nine members of the Detroit elite responded by raising an initial $10,000, and another $12,000 came from a larger pool of Detroit donors when the project required it.[31] Although the observatory itself was located in Ann Arbor, it was named the Detroit Observatory as a monument to the civic community that had funded the project. In New York, in 1854–1855, Hamilton College's fund-raising wizard, Professor Charles Avery, raised $20,000 by public subscription for an observatory building and a large 13-inch refractor. Avery also managed to convince railroad promoter and property developer Edwin Litchfield to provide the rarity of an ample endowment, $30,000, for the honor of naming the institution the Litchfield Observatory.[32] All of these examples show the geographic range across which the public-subscription model was applied to the financing of sizable observatories.

Public subscription was, however, not the only way in which communities organized to fund large observatories—nor was the city the only locus of community support. The extravagant Barnard Observatory at the University of Mississippi, which was designed to house the largest telescope in the world, managed to raise its funding directly from the government of the state of Mississippi. As with so many university observatory projects, the Barnard Observatory was the initiative of its president and chancellor Frederick August Porter Barnard. Barnard, whose grandfather had been John Quincy Adams's secretary of war, developed an early passion for science and astronomy and had been appointed professor of natural philosophy at the University of Alabama at the age of twenty-eight.[33] He had been the driving force behind the founding of the University of Alabama observatory in 1844, with its large central dome, transit room, and Simms equatorial telescope, the latter costing an estimated $4,000.[34] He became the first president of the University of Mississippi amid concerns over the tide of young men flowing to the North for their education, and he assured the trustees and the state that he could reverse the flow.[35]

To bring the level of the University of Mississippi up to the best in the world, Barnard convinced the Mississippi elite to allow a special appropriation by the state legislature in 1856 for a $100,000 revitalization project.[36] Astronomy was a crucial part of his plan, and a large portion of the funds would be used to signal that Mississippi was second to none by building the largest telescope in the world. Barnard contracted with the American firm Alvan Clark & Sons, for an unprecedented 18.5-inch objective lens, commissioned a massive 11-foot-diameter orrery, and imported thirty-six celestial and terrestrial globes from Malby & Sons in London.[37] As at Albany, a grand edifice was built before the telescope arrived, this one copying exactly the layout and appearance of nothing less than the imperial Pulkovo Observatory in Russia. So fundamental was the observatory to Barnard's vision for the university that he moved his entire family into lodgings within its walls as a symbol of what was to be the new campus's spiritual and intellectual center. However, before the telescope could be delivered, the Civil War commenced and the state reappropriated all unspent funds for use in the war effort, forcing the university to default on its payment for the refractor. Although the plan never came to full fruition, it is another strong example of observatories being used for community signaling, in this case through state funding support rather than through citizen subscription.

The importance of the civic observatories in astronomy would soon decline, however, and the stories of the two last significant examples of the type—the Allegheny Observatory and the Dearborn Observatory—point to the reasons why. The Allegheny Observatory was established by a shareholder corporation, similar to Mitchel's organization for the Cincinnati Observatory. Its establishment was driven largely by an interest in exploratory astronomy that had been engendered among the elite in Pittsburgh thanks to the efforts of local educator Lewis Bradley. In February 1859, Bradley called together four men who had expressed an interest in buying a large telescope after viewing the heavens with Bradley through his small refractor. These four consisted of a newspaper publisher, a wholesale shoe merchant, and two leading bankers, one of whom would later become a congressman. By March, through their connections, these individuals had enlisted the support of an additional sixteen, all of whom pledged $100 for membership in the Allegheny Telescope Association, which would grant them the privilege of personal access to the new telescope.[38] Unlike the Cincinnati Observatory, which had grown out of a relatively broad-based interest in astronomy, or the Dudley Observatory, which had arisen largely from a desire on the part of the city leaders to signal the cultural prowess of Albany, the Allegheny Observatory's origin was a handful of members of the Pittsburgh elite who wanted an exclusive location from which they could observe the heavens. No research was to be done at the observatory; the purpose was simply the personal exploration of the Moon and planets by members of the association. Other than a large refractor, which would be one of the largest in the world, there was to be no scientific instrumentation at the observatory, a testament to the singular interest in the personal exploration of the heavens that drove the observatory's founding.

Although this type of enthusiasm allowed for rapid development, it also made the observatory's financial situation particularly vulnerable. The initial group secured a lecture series by Mitchel that increased interest and subscriptions to sixty-six incorporators. This ultimately led to some $14,000 being promised in subscriptions by November 1859—a sum that allowed the group to contract with Henry Fitz for a large thirteen-inch refractor.[39] An act to incorporate the observatory was passed by the Pennsylvania legislature, and an observatory was built in anticipation of the delivery of the telescope in September 1860. When the Civil War started in April 1861, however, the telescope had still not been delivered. With the onset of the war, the men of industry who had supported the project, none of whom had a prior interest in astronomy, turned to the more pressing and profitable matters of war provision. When the observatory was finally dedicated in January 1862, only a small group of seventeen arrived for the proceedings. The war had also greatly affected the ability and willingness of the individuals to make good on their promised subscriptions, and the organization was forced to go some $12,000 in debt. By 1867, lack of interest and mounting debt caused the Allegheny Observatory to be subsumed into the Western University of Pennsylvania, now the University of Pittsburgh, which had long coveted the observatory's prestige and extensive land holdings.

Unlike the community engagement that had supported the initial observatory, this reestablishment as part of the Western University of Pennsylvania would be principally enabled by a single man, William Thaw, a steamboat and railroad magnate who had been one of the observatory's original incorporators. Thaw provided some $20,000 of the $32,000 required to pay off the Allegheny Telescope Association's debts and to establish an endowment that would allow for research at the observatory. This endowment led to the hiring of the influential astrophysicist and later pioneer of aeronautics Samuel Pierpont Langley as its new director.[40] After the association of gentlemen had failed to properly finance their endeavor, this new funding model—reliance on a single prominent donor—became the mainstay for the observatory. It was also to become the new funding model for American astronomy in general. Thaw and his son William Thaw Jr. would continue to generously support the observatory and astronomy in Pittsburgh more generally. The culmination of their support was the New Allegheny Observatory, begun in 1900 and completed in 1912 at a total cost of $300,000, with its $125,000, thirty-inch Thaw Memorial Refractor—the third largest in the world at the time.[41] Astronomy in Pittsburgh, which had begun in the tradition of the civic observatories, ultimately had to rely on a single, major benefactor for its growth.

The last of the civic observatories would take root in Chicago in the midst of the Civil War. Although our understanding of the observatory's history is greatly hindered by the loss of its records in the Great Chicago Fire of 1871, we can stitch together a general outline of its founding. In November 1862, a man identified in the literature only as "Professor M. R. Forey" came to the University of Chicago looking to sell a large refractor made by American telescope maker Henry Fitz.[42] Although the nature of his interest in astronomy remains a mystery, it seems likely that the "Professor M. R. Forey" was the former president of the Chowan Baptist Female Institute in North Carolina and Baptist minister Reverend Martin Rudolph Forey—an individual described in a 1906 history of Hertford County as "a Christlike man, of great literary culture, and of wonderful energy and business sagacity."[43] After meetings with the university president and prominent university patrons, arrangements were made for Forey to give a lecture called "The Sidereal Heavens" in Bryan

Hall, the largest auditorium in Chicago. The objective was to generate support for a university-based observatory. Immediately after the address, a small, tight-knit group of civic leaders, led by banker and railroad owner J. Young Scammon, founded the Chicago Astronomical Society with the objective of building an observatory in Chicago. Membership in the society would cost $100 and would entitle members to visit the observatory once a week with their families; a life directorship of the society could be secured for a $500 donation.[44] Although the society's genesis had its origins in Forey's offer of a Fitz refractor, the investigations of the newly enthusiastic society members soon led them to discover an even more enticing prize—the large 18.5-inch Clark objective lens, once destined for the University of Mississippi, which had recently become available with the onset of the Civil War.

Seeing the chance to secure for Chicago the largest refractor in the world, cash donations for an initial installment of $1,500 were quickly raised. The new Astronomical Society's secretary, Illinois State District Attorney Thomas Hoyne, hurried to Boston in late January to contract with Clark & Sons for the telescope before Harvard, which was also interested in the refractor, could close the deal.[45] The $18,187 cost of the lens and mounting was raised by public subscription, likely with significant contributions from the initial group of civic leaders.[46] A former Chicago mayor donated $7,400 to import a German transit circle, but it was the business magnate Scammon himself who bore the majority of the cost, paying the roughly $30,000 required for the observatory dome and tower that would crown Douglas Hall, the University of Chicago's new main building.[47] This gave Scammon naming rights for the observatory, which thus became the Dearborn Observatory, after his wife's maiden name.

The civic leaders of Chicago, through quick action and by leveraging the initial investment of Mississippi, had secured for their home something no other America city could claim: possession of the most powerful instrument of space exploration on the planet. Although we lack the detailed documentation on the driving motivations of the individuals involved, it nonetheless seems clear that the Dearborn Observatory was the product of a civic community, proud of their city and eager to signal its arrival on the world stage. As it had been at other civic observatories, it was a mark of pride for the community to have a technological capacity that would allow them to see the wonders of the heavens for themselves. As had been the case elsewhere, they organized themselves in significant numbers to achieve this, with the most powerful community members inspired by one another, and by the response of the general public, to put their resources behind the project. This signaling motive would remain strong in the founding of later nineteenth-century observatories. The source of funding, however, would begin to shift. The primary funding role played by the business magnate Scammon in the case of the Dearborn Observatory would become even more pronounced in the later observatory era. As the Gilded Age concentrated increasingly vast wealth into the hands of a few individuals, the scope expanded for observatories to rely largely on a sole patron or founders. It would no longer be civic communities but rather single, wealthy individuals that would be the mainstay of support. The era of the super-rich patron, and the most expensive observatories, was about to begin.

Prior to considering the surge in large, privately financed observatories funded principally by single individuals, a couple of general observations are worth noting. First, as the nineteenth century progressed, professional American astronomers moved further away from the traditional mainstay of positional astronomy, and increasingly focused on what became known as "the new astronomy"—an approach to investigating the heavens that focused more on the physical nature of the objects in space and that ultimately would lead to the development of astrophysics and the planetary sciences. Popular American interest in astronomy had long been rooted in the excitement and awe attendant with examining heavenly bodies and planetary worlds. Professional astronomers, however, had largely maintained their focus on less-speculative research efforts—namely positional astronomy. The increasing resolution of telescopes and new techniques such as photography and spectral analysis, however, gave astronomers new tools with which to conduct investigations that could go beyond speculation and which began to provide answers to long-standing scientific questions about the physical universe. These investigations and their attendant discoveries, which often had great appeal to the popular imagination and thus received significant media coverage, increased the prestige value of observatories. These scientific discoveries were often heralded as major milestones in human knowledge, thus more closely aligning the interests of the scientists pursuing them and patrons interested in establishing a legacy. At the same time, the shift away from the practical applications of positional astronomy—such as timekeeping, surveying, and longitude determination—narrowed the spectrum of functionality that observatories could seek to claim. These factors decreased the utilitarian appeal of astronomical observatories, but at the same time they increased their signaling value as credible signals of wealth and of commitment to lofty intellectual pursuits.

A second relevant trend was the increasing concentration of wealth in America over the course of the nineteenth century. America in general was undergoing rapid economic growth in the nineteenth century, but the wealth of the richest Americans was growing even faster. By 1892, a *New-York Tribune* survey could report that there were over 4,047 millionaires in the United States, equivalent to there being 4,047 individuals with over $1.1 billion in equivalent 2015 GDP-ratio terms—an order of magnitude more than the number of people with control over a comparable share of the nation's resources in 2015.[48] We currently lack the data to demonstrate conclusively that income inequality is positively correlated with private expenditures on space exploration over long periods of time. It does seems a likely hypothesis, however, given that the largest private expenditures on American astronomical observatories came at the end of the nineteenth century and the beginning of the twentieth, and we have seen significant private expenditures on spaceflight capabilities in the beginning of the twenty-first century—times when income inequality in America has been at its highest.

Although the trend of single individuals endowing large observatories peaked in the late nineteenth and early twentieth centuries, the trend can be seen throughout the history of observatory funding in America. There is no abrupt or complete moment of transition in the nineteenth century— it is instead a general trend of increasingly important singular observatory donations. Many of the civic observatories already discussed were enabled by large donations from single benefactors such as Scammon, Thaw, Dudley, and Sears. The role of individual patrons in the establishment of astronomical observatories would increase and diversify into new contexts as the century progressed. From the numerous, singularly

endowed university observatories to the personal observatories of Rutherford and Lowell—and to the independent, monumental institutions
of Lick and Mount Wilson—the private wealth of individuals became the
principal source of funding for astronomical observatories in the nineteenth and early twentieth centuries.

It is in the context of university observatories that the prominence of
individual donors can first be seen, including in the civic observatories,
which tended to be embedded within a university context. Scammon and
Sears provided funds specifically for university-situated observatories at
Chicago and Harvard, and Thaw's major contributions came only once
the Allegheny Observatory was to be transferred to the Western University
of Pennsylvania. Individual patronage has a long history at universities
in general, as well as at the early American university observatories, such
as Clark's donation to Yale, Jenkins's patronage of Georgetown, and Phillips's bequest to the Harvard College Observatory. This trend continued
to fund numerous small-to-moderate-size university observatories throughout the century. In 1852, for example, a $7,000 donation from Boston
physician George Shattuck provided the principal funding for the Dartmouth College Observatory in New Hampshire, with an additional
$4,000 coming from the college.[49] In the mid-1860s, astronomy was a
priority for the wealthy brewer Matthew Vassar, who constructed, as the
first building on his new Vassar College campus, an observatory with a
large Fitz objective lens—at a total cost of more than $14,000—to en-
tice the famous comet discoverer and astronomer Maria Mitchell to join
the college as its first faculty member.[50] It was during the decade after the
Civil War, however, that the rise of patron-funded observatories truly
took hold.

The story of one of the most important patron-initiated observatories,
the McCormick Observatory at the University of Virginia, is intimately
related to the story of the last of the civic observatories, the Dearborn
Observatory. The observatory's founding was initiated by Leander McCormick, the youngest of the three brothers responsible for the McCormick
Harvesting Machine Company—the leading producer of American farm
machinery during the agriculture boom of the late nineteenth century.
Part of the motivation for the project may have been competition between
Leander and his older, more-famous brother, Cyrus. It has also been argued,

however, that the rivalry at the heart of the McCormick Observatory was not a sibling one, but rather the long-standing American rivalry between North and South.[51] The McCormicks were pro-slavery Southerners living in Chicago when the founders of the Dearborn Observatory bought, during the middle of the Civil War, the 18.5-inch lens originally meant for the University of Mississippi. The world's largest telescope, originally destined for the South, was now installed with much publicity in the North. Thomas Williams argues that the patriotic Southerner McCormick saw this as yet another symbol of the South's defeat and resolved to donate his own "world's largest telescope" to the University of Virginia so that it "could have a powerfully restorative effect on morale in the state, and might, if properly exploited, lead to years of newsworthy honor for the institution and for the state itself."[52] Either way, it seems likely that it was signaling and rivalry, whether at the interpersonal or interregional level, that provided a familiar spur to McCormick's major investment in astronomy.

Although the motivation was familiar, never before had a single individual in America attempted to personally fund a telescope of such magnitude in its entirety and to personally direct the terms of the observatory. In 1870, as a first step in the project, he ordered a twenty-six-inch refractor from Alvan Clark & Sons for $42,000 and discussed the project with General Robert E. Lee, then president of Washington College, whom McCormick expected to provide an observatory building and endowment for "his" telescope from the university funds.[53] McCormick later envisioned providing for the observatory more generally with an unprecedented planned contribution of $100,000–$150,000. Despite these ambitious plans, however, the funding of the McCormick Observatory eventually required more than a single patron effort. The Chicago Fire of 1871 and the financial panic of 1873–1874 hindered McCormick's ability to proceed with the project. Its completion was only assured in 1881 when the University of Virginia raised some $75,000 in donations from William H. Vanderbilt and university alumni, and also accepted McCormick's donation of the twenty-six-inch objective lens along with an additional $18,000 to fund the building.[54] Although McCormick ended up funding less than half of the observatory, his early ambition of personally funding the world's largest telescope was a sign of the individualistic nature of the developments to come.

The early 1870s saw the founding of a number of large university observatories that were made possible principally by the desires of single patrons. After an abortive, earlier attempt to build an observatory at the College of New Jersey—now Princeton University—an astronomical observatory topped the college's postwar wish list. Brigadier General Nathaniel Halsted, a prominent New Jersey dry-goods merchant, had promised to contribute to the observatory campaign before the war and made good on his promise with an initial gift of a $10,000 bond, a donation that he later increased to the entire cost of the building—$60,000 when completed in 1872.[55] This substantial donation meant the observatory building's completion preceded, by almost a full decade, the installation of its twenty-three-inch Clark telescope—the $32,000 cost of which was raised by subscription.[56] A similar generosity took root at Yale, where local New Haven rifle baron Oliver Winchester gave the university land valued at $100,000 in 1871.[57] It was hoped, by both donor and university, that the ongoing post–Civil War land boom would raise the value of the property to some $500,000—an amount that would allow the Winchester Observatory to possess the world's largest telescope and the most impressive of observatory buildings, the designs for which show an extravagantly decorative architecture. The depression of the mid-1870s, however, burst the Connecticut property bubble, and the land never reached the hoped-for value—although it did allow for a sizable observatory with a twenty-eight-inch Clark refractor. Signaling and legacy motives had led to the establishment of the Winchester and Halsted Observatories; with no financial support for astronomers included, scientific interest was at best a marginal motivation.

The Morrison Observatory highlights the role that personal interests in astronomy could play in these decisions, however. The observatory was founded in 1874 with two $50,000 gifts—for the observatory and for an endowment for its operation—coming from plantation heiress Berenice Morrison.[58] Mrs. Morrison's decision to donate the entirety of the required funds for an observatory at the Pritchett School Institute has been attributed to her viewing of the Great Comet of 1874 with Carr Pritchett, the founder of the institute.[59] However, her own autobiographical manuscript, *Plantation Life in Missouri,* makes it clear that Mrs. Morrison also had a love of the sky that extended from childhood: "I could plunge my eyes into the stars—I felt caught up by a strange power, an unspeakable longing possessed me, a mystery enfolded me and my heart, my childish but untrammeled heart, beat with a great happiness. I would resolve to gaze upon the marvelous heavens all night, feeling myself an intimate part of these wonders."[60] This intrinsic interest in astronomy may have been part of the reason why Morrison, unlike many of her fellow observatory patrons, also included funds for a research endowment. Personal interest could thus be just as important a driver for the observatory patrons as a desire for legacy. It would be both of these motives together, however, that would drive the most resource-intensive space exploration project of the nineteenth century.

The Lick Observatory was a milestone in the development of the physical and cultural infrastructure of American space exploration. Although single individuals had wholly endowed observatories before, they had never done so on the scale of Lick. James Lick had made his fortune in real estate during the California gold rush and, as his health began to deteriorate, he decided that he wanted to build a monument by which he would be remembered. He is said to have considered a number of monumental structures, including gigantic statues of his parents that would overlook the San Francisco Bay, and a giant pyramid to be erected in the middle of the city. Through discussion with a number of astronomers, however, including George Madeira and George Davidson, he was convinced that the most spectacular monument he could achieve would be to build the world's largest telescope and to enable it to pursue prestigious, legacy-creating, scientific research—and then to be buried underneath it, which he was in 1887. Although he was unquestionably motivated by the legacy that such an observatory would leave—he had initially insisted that the observatory be built in downtown San Francisco, where it could be seen—the farsighted Lick also had an appreciation for space exploration that went beyond even that of most contemporary astronomers. He once confided to a long-time friend that he thought that someday man would walk on the Moon and that "We will know the secrets of the spheres and it will be as common for man to take an inter-orbital trip into space as it is for you or me to walk down Montgomery Street."[61] With a vision of such a future, it seems unlikely that his selection of an observatory over other possible monuments was motivated solely by signaling considerations. Lick was evidently a man who had a direct interest in the exploration of the heavens. It was this personal interest that, when combined with his extreme wealth and desire for legacy, resulted in the establishment of the world's most advanced astronomical observatory.

Lick's interest extended to personally directing the earliest phases of the observatory's development. One of his most significant and telling decisions was to make the Lick Observatory the first mountaintop observatory, selecting the site of Mount Hamilton, around forty kilometers from his adopted home of San Jose. This decision sacrificed some of the public accessibility of the monument, but it resulted in improved seeing conditions that led to major advances in astronomical photography and research. Lick also sacked the original board of trustees when they failed to operate as he had hoped, and he personally lobbied officials in Santa Clara County to pay the $73,000 for the construction of a road up to Mount Hamilton.[62] At Lick's death in 1876, his will left an unprecedented $700,000, 17.5 percent of his nearly $4-million estate, for the establishment of an observatory in his name that would house the world's largest telescope—the cost of which represented a share of U.S. GDP equivalent to roughly $1.3 billion in 2015.[63] The Lick Observatory, which became one of the central institutions of American astronomy and astrophysics in the decades after its construction, is an example of the significant resources that can be mustered by the private sector for the exploration of space, based solely on an individual's personal interest and desire for legacy.

Although there would be no observatories of comparable magnitude to Lick's until those built by George Ellery Hale, more moderate, but nonetheless economically significant founder observatories continued as the norm until the Second World War. Most of these projects can be broadly categorized as being motivated by either signaling or intrinsic interest, although there are also interesting subcategories that are worth noting. For example, outright self-promotion and commercial advertising was part of the direct impetus for Hubert Warner's decision to found his observatory. As his native city of Rochester swelled with pride over its newly famous comet-discovering astronomer Lewis Swift, Warner, a maker and promoter of patented medicines, including "Warner's Safe Kidney and Liver Cure," decided in 1880 to build a downtown observatory with a sixteen-inch refractor at a cost of $100,000.[64] When the building was complete, it was an ostentatious example of Gilded Age Gothic architecture and was designed to achieve Warner's dual aims of signaling his status in the community and attracting attention to himself and his Safe Pills. In 1881, he announced a "Warner Safe Remedy Prize" for the first American discoverer of each new comet, valued at $200—equivalent in 2015 PWC-ratio terms to $55,500. He subsequently increased the value of the prizes to $1,000, equivalent to roughly $240,000 by the same metric.[65] To maximize its publicity, Warner made the observatory the first in the world to be permanently open to the public, and it soon became a genuine tourist attraction for the city of Rochester. The prestige and publicity associated with astronomy could be shrewdly exploited for advertising and commercial gain even in the nineteenth century.

Another interesting case is that of the Mount Lowe Observatory, to which both Lewis Swift and the sixteen-inch telescope from the Warner Observatory later moved. The Mount Lowe Observatory was another commercial endeavor—in this case, an attempt to capitalize on the public appeal of astronomy in the context of a commercial tourist resort. After the financial panic of 1893, Warner was no longer able to support the Rochester observatory and Swift accepted an offer from eccentric aeronaut, inventor, and businessman Thaddeus Lowe to move to California and become the director of a new observatory. The observatory was part of the Mount Lowe Railway, a sprawling collection of hotels, zoos, restaurants, and other tourist attractions that Lowe had built along the all-electric streetcar line that ran from Pasadena up the San Gabriel Mountains. The Mount Lowe Railway was one of the most popular attractions in the Los Angeles area in the late nineteenth and early twentieth centuries and was a forerunner of Disneyland-style family entertainment. The observatory, while only one of the many attractions, was an important draw. It featured prominently in the brochure and was open for public exploration of the heavens three nights a week. Even in the late nineteenth century, individuals within the leisure and entertainment industries understood the value of space exploration in attracting the attention, and the coinage, of the general public.[66]

Another interesting subcategory of founder observatories were those that were funded personally by state governors, such as the Washburn and Ladd Observatories in Wisconsin and Rhode Island. In 1876, after the regents of the University of Wisconsin had complained that no university worthy of the name could be without an astronomical observatory, the Wisconsin legislature passed an act that provided an annual grant of $3,000 for university astronomical work, on condition that an observatory could be built without state funds.[67] Cadwallader Washburn, former Wisconsin governor, three-term congressman, and businessman, personally endowed the observatory with the requisite funds amounting to $65,000, with specific instructions that its refractor be larger than Harvard's—a feat that it achieved by commissioning a 15.6-inch Clark objective lens to beat out Harvard's 15-inch Fraunhofer refractor.[68] In Rhode Island, it was a sitting first-term governor, Herbert W. Ladd, who promised to personally donate the funds for an observatory at Brown University after attending an alumni dinner in 1889. With a cost increase, mostly due to decorative elements in the observatory's architecture, the facility was completed in 1891 at a cost of $30,000, shortly after the start of Ladd's second term as governor.[69] These observatories represent the continuation of an interesting trend: state-level signaling, which had previously been focused on the legislative bodies, such as in Pennsylvania and Alabama, was now taking root at the executive level in the era of founder observatories. They also show that politicians did not ignore the public relations benefits that patronage of an astronomical observatory could provide.

Similarly, although broad-based funding for civic observatories had declined, civic sentiments still inspired wealthy individuals in the new boomtowns of Denver and Los Angeles to fund astronomical observatories. Like Cincinnati and Albany at the time of their observatory projects, Denver was undergoing rapid growth during the Colorado silver boom of the 1880s. As James Lick had during the California gold rush, Humphrey Chamberlin had arrived in Denver by chance at the start of a massive boom and experienced a rapid rise to riches with his successful property speculations. Unlike the more reclusive Lick, however, Chamberlin was an active figure in the Denver community, owning major shares in numerous local companies, founding the Denver Savings Bank, and becoming president of Denver's Young Men's Christian Association and the Denver Chamber of Commerce and Board of Trade. Although he was the patron of a number of local philanthropic endeavors, his most expensive project was the Chamberlin Observatory of the University of Denver. The project was initiated in 1888 and, with a twenty-inch Clark refractor, cost Chamberlin a total of some $56,000.[70] Although the 1893 silver-price crash destroyed Chamberlin's wealth and left the observatory without an endowment, the large telescope and imposing edifice—which had been copied wholesale from the $65,000 Goodsell Observatory begun at Carleton College in Minnesota at the same time—ensured that it remained an important symbol of civic pride.[71]

The Griffith Observatory of Los Angeles also had a predominantly civic context. Griffith J. Griffith, who had settled in Los Angeles after making his fortune in mining, had granted his adopted city over three thousand acres for use as a city park in 1896. In his will, he left his estate to the city, with instructions that it be used to fund a large observatory in the park for civic enjoyment and education. When the estate was settled after his death in 1919, some $225,000 was allocated for the observatory.[72] Although civic communities no longer made collective attempts to fund major observatories, individual philanthropists continued the tradition of civic-minded astronomy. As with the earlier civic observatories, however, this interest did not extend to funding professional astronomers to use the expensive telescopes to perform scientific research.

There was an intrinsic interest in scientific research in astronomy as well, however, and in the cases where this interest was combined with personal wealth the results could be scientifically productive observatories. The oldest of the funding traditions in American astronomy, that of individuals funding their own astronomical facilities for their personal use, also took on a new scale in the mid- to late nineteenth century. Starting with the first telescope in America, imported in 1660 to satisfy John Winthrop Jr.'s passion for astronomy, wealthy individuals indulging in their own explorations of the heavens had played an important role in the spread of astronomical observatories across the nation. A full listing of the innumerable small, personal telescopes and observatories is neither possible nor necessary here. Some of the more significant early efforts are, however, worth noting. One particularly interesting astronomy enthusiast was John Jackson, a Quaker minister and estate owner in Darby, Pennsylvania, who, in 1845, added a large observatory tower and an imported 6.3-inch Merz equatorial telescope to the family home—which was also a girls' boarding school—at a total cost of $4,000.[73] While there were a number of other astronomy enthusiasts on country estates throughout the nation, it was from within the wealthier social circles of America's growing cities that the American "Grand Amateurs"—individuals who personally funded significant astronomical facilities and establishments for their own personal use—emerged.

Although the Grand Amateur tradition did not quite reach the extent in America that it did in England, there were nonetheless a number of notable amateur contributions to American astronomy.[74] Lewis Morris Rutherfurd was one such self-funded astronomer, who conducted pioneering work in astrophotography and stellar spectroscopy and who also provided important support for American telescope maker Henry Fitz.[75] Rutherfurd was fortunate enough to be born into a wealthy New York family and to marry into an even wealthier one. This allowed him to retire from his nascent career as a lawyer and dedicate himself fully to astronomy. Starting in 1856, he built up his personal observatory, situated on the corner of Second Avenue and East Eleventh Street in New York, into one of the best equipped in the nation, with numerous astronomical instruments, including a $2,200 nine-inch Fitz refractor and eventually one of thirteen inches.[76] Henry Draper was another New York Grand Amateur, whose research in astronomy was significantly assisted by his marriage to a wealthy heiress.[77] Although he had long devoted his leisure to astronomy, his marriage provided him with the means to build a world-class astrophysical laboratory and to manufacture his own twenty-eight-inch reflector—resources that enabled him to be the first to successfully photograph an astronomical spectrum, that of the star Vega in 1872. There were a number of other amateurs as well, of lesser fame perhaps, but who nonetheless represented significant private expenditure on astronomy. Robert Van Arsdale, for example, was an avid comet hunter from his home observatory in Newark, New Jersey, with a $1,125 clockwork driven Fitz telescope purchased in 1850.[78] A nine-inch Fitz refractor was the centerpiece of William Van Duzee's private observatory—built in Buffalo in 1851—the base of which was fixed twelve feet below ground level so that passing carriages would not disturb the telescope. At the same time in New York, a Brooklyn rooftop observatory housed the twelve-inch Fitz refractor of Jacob Campbell. With the development of an American telescope manufacturing industry—led by Amasa Holcomb, Henry Fitz, Alvan Clark & Sons, and John Brashear—individuals were increasingly able to "found" their own observatories right on their own properties.[79] None, however, would do so on the scale of the grandest of the American Grand Amateurs, Percival Lowell.

At the turn of the century, in the same decade that U.S. rocketry pioneer Robert Goddard made his personal commitment to the development of spaceflight, Percival Lowell made his decision to dedicate himself to the remote exploration of Mars. Unlike Goddard, however, Lowell was wealthy enough to fund his efforts himself. The son of a wealthy Boston family, Lowell had developed a passion for astronomy at a young age and had traveled extensively in Japan and Korea, gaining a love of exotic adventure. Upon return from his Asian travels in 1893, he learned of the retirement of Giovanni Schiaparelli, the observer of the "canals" of Mars. Lowell decided that he would take up Schiaparelli's mantle and dedicated his life to further exploring the red planet.[80] He soon began construction of the largest personal observatory in America, the Lowell Observatory, to pursue his passion. From his observations, he became increasingly convinced that he could see evidence of artificial canals on Mars, which he believed were a sign of intelligent life. His popular accounts of life on Mars, *Mars and Its Canals* and *Mars as the Abode of Life* in 1906 and 1908, greatly excited the general public and influenced the cultural milieu from which the early rocket societies emerged.[81] In time, the Lowell Observatory would become one of the critical research observatories in American astronomy, with its research programs resulting in the discovery of Pluto in 1930.

Thanks to the work of David Strauss, we have a reasonable knowledge of the history of Lowell's finances and the funding of the observatory. At the start of the observatory's development in 1894, Lowell had a sizable annual income of $25,000.[82] Although Strauss was unable to find financial records for the actual construction of the observatory, he was able to uncover an expenditure of $2,300 for the four months of expeditions and planning that led to the site selection. In 1896, there are records for the purchase of the twenty-four-inch Clark refractor, the fourth largest in the United States at the time, at a cost of $20,000. The most interesting financial information, however, relates to the operating expenses of the observatory. These results provide valuable insight into the magnitude of Lowell's passion for space exploration and his willingness and ability to fund it.

Changes in the annual funding of Lowell's observatory reveal the relative income inelasticity of his demand for space exploration. In 1897, Lowell instructed his assistant that the total annual expenses were not to exceed $6,000 per year.[83] Lowell was thus willing to spend around 25 percent of his income annually on his observatory at that time. When his father died in 1901, Lowell's inheritance meant that his annual income grew from $25,000 to $100,000.[84] At the same time, his average annual expenditure on the observatory grew to a little more than $10,000 over the period from 1903 to 1908. This means that the relative share of Lowell's annual income that he decided to spend on his observatory decreased from around 25 percent to around 10 percent. Lowell's expenditure pattern thus provides an interesting, though anecdotal, example of the income inelasticity of an individual's demand for astronomical investigation. Lowell was determined to fund an observatory of significant scale and did so even when it represented a major portion of his wealth. He increased his expenditure as his wealth increased, but he did so at a decreasing rate. In this crude sense, Lowell's personal demand for the exploration of the heavens was inelastic over time.

The Perkins Observatory was the result of a similar personal dedication to astronomy and the continuation of a tradition that dates back to the earliest college observatories—that of faculty providing the resources to establish observatories. During the Civil War, Hiram Perkins, a professor of mathematics and astronomy at Ohio Wesleyan University, had wanted to join the Union Army but was refused on account of his abnormally large height and low weight. Still committed to supporting the Union, he decided to devote his family farm to raising pigs to provision the army with salt pork. As a frugal-living man, his pork sales during the war meant that he managed to amass a sizable fortune. After he eventually retired from the university in 1907, he spent the next fifteen years of his life planning the construction of a sixty-one-inch reflector, which would become the first large telescope mirror cast in America and the third largest reflecting telescope in the world. He was motivated in part, as *The Christian Advocate* reported in 1922, by "an ideal that dominated his whole life—the belief that the study of astronomy outclasses all other studies in teaching the majesty and power of God and in inculcating principles of true religion."[85] Starting in 1923, Perkins spent some $288,000 on the reflector, observatory, and telescope, while his sister provided $91,000 for an endowment.[86] Despite its world-class facilities, however, Ohio's cloudy weather, low elevation, and light pollution from nearby Delaware and Columbus limited the observatory's productivity, as did Perkins's death before its completion in 1931. Nonetheless, the Perkins Observatory shows that, even in the era of the large founder observatories of the early twentieth century, the same combination of religious sentiment, college fealty, and personal interest that had motivated the foundation of American observatories almost one hundred years earlier was still a potent force in the patronage of astronomy.

The last significant founder observatory of the early twentieth century, the McDonald Observatory, was motivated by that old friend of astronomy—the concern of wealthy men for a lasting legacy. William Johnson McDonald was a prosperous Texas banker who, to the surprise of the University of Texas and the consternation of his family, decided to bequeath the bulk of his fortune to the university for the construction of a large astronomical observatory. McDonald had possessed a small telescope, and astronomy was one of his many hobbies, but the extent of his gift to astronomy was entirely unexpected. When he died in 1926, his will specified that over $1 million of his estate, valued at $1.26 million, should be given to the university to build and endow an observatory in his name.[87] McDonald's relatives and extended family contested the will, however, and after a three-year court battle, with no end in sight, the university settled with the contestants out of court, leaving approximately $840,000 for the observatory.[88]

The development of the observatory—which would house an eighty-two-inch reflecting telescope, the second largest in the world at the time of its dedication in 1939—saw the beginnings of a number of trends that would be ingrained into the business of astronomical research and space exploration throughout the rest of the twentieth century. The type of multi-university collaboration that would become the model for large observatories later in the century was presaged by a unique partnership for the design and subsequent operation of the observatory. Lacking a strong faculty in astronomy, the University of Texas decided to partner with the Yerkes Observatory at Lake Geneva, Wisconsin, which had an experienced faculty but limited funding. The increasing complexity of astronomical technology and a desire to keep oversight simple led the University of Texas to select a prime contractor, the Warner & Swasey Company, to execute the entire project for a firm fixed price—a new procurement model for an astronomical observatory and one that, for the same reasons, would later become the preferred method used by the private sector to purchase spacecraft. Although the McDonald Observatory shared much in common with modern projects of space exploration, it also marked the end of an era. William McDonald was the last individual to personally endow a large observatory in America until the 1990s, drawing to a close a golden age of founder observatories. To complete the analysis of the organization and funding of large observatories, however, we also need to examine the unique narrative of the man who would become America's greatest observatory promoter, George Ellery Hale.

Hale is without equal in the history of nineteenth- and early-twentieth-century space exploration. His three successive "world's largest telescopes" were at the pinnacle of American astronomy from the 1890s until the 1990s, when his last and most important work, the two-hundred-inch reflecting telescope at the Palomar Observatory, was finally surpassed in resolving power over forty years after its completion. Although a respected scientist and capable instrument builder, Hale's real strength was as the initiator, principal organizer, and unparalleled fund-raiser for the mammoth projects that he pursued. Over his career as a fund-raiser, Hale tapped a diverse mix of sources, which shifted as his reputation grew and as the landscape of American philanthropy began to change in the early twentieth century. After first convincing his father to bankroll the construction of an impressive solar observatory in the backyard of the family home, Hale moved on to personally raise observatory funds from some of the wealthiest, most powerful people in America and to pioneer the relationship between large observatories and the then newly emerging large philanthropic foundations. Hale's story is a microcosm of the broader funding narrative of American astronomy. It highlights the important role of signaling and personal interest in observatory funding, but it also un-

derlines that it is the often passionate and dedicated efforts of the virtuoso individual that provide the motivating force. By examining Hale's
correspondence with his patrons in the George Ellery Hale Papers at the
California Institute of Technology Archives, and building on previous
Hale scholarship, it is possible to analyze the personal strategies and underlying motivations that allowed Hale to become the preeminent astronomical entrepreneur.

The story of Hale's early dedication to astronomy has much in common
with what we see in the stories of the early rocket pioneers. Like Goddard,
Hermann Oberth, and Konstantin Tsiolkovsky, Hale was a sickly child
who found solace and purpose at an early age in books, study, and
thought. As with the rocket pioneers, his career decision can be attributed to a sort of "conversion" associated with spaceflight fiction. Hale had
been a young boy of many interests until he read Jules Verne's *From the
Earth to the Moon,* after which his passion began to bend more and more
toward astronomy.[89] As he wrote in his unpublished notes for an autobiography, "Books of adventure were read by the score, especially Jules
Verne's 'Mysterious Island,' 'Twenty Thousand Leagues under the Sea'
and 'A trip to the Moon.' The gigantic telescope on a peak of the Rocky
Mountains, into which 'the unfortunate man disappeared' while watching the projectile on its way to the moon, especially struck my fancy."[90]
The inspiration of Jules Verne's adventures was deeply ingrained, and Hale
portrayed astronomy as an exploration and a grand adventure in his public
and private writing. His "Recollections of Childhood," written near the
end of his career in 1935, starts off in a rollicking tone: " 'Adventures'? you
say, as you scan these pages. Certainly not, if looked at from the standpoint
of an early Californian pioneer or a soldier in the front trenches of the war.
But if adventures can happen anywhere and consist of all sorts of events,
some of them involving heavy risks, of one kind or another, I have had quite
enough. After all, an astronomer may see strange worlds and wander far
from the orthodox path in half a century's experience at home and
abroad."[91] After his experience with Verne, astronomical adventure promised in "strange worlds" became the young George Hale's overriding interest, and he became, in his own mind, a self-starting space explorer. Unlike
the similarly inspired early rocket pioneers, however, Hale started life as
the only son of a family that could directly fund his ambitious projects.

Hale's father, William, made his fortune in the Chicago elevator business during the postfire rebuilding boom. He would provide major financial support for his son's pursuits throughout his life, starting with the purchase of young George's first telescope. As Hale's enthusiasm for science grew, his father outfitted the attic of the family's new mansion in Kenwood as a laboratory and purchased for his son the finest in photographic equipment, heliostats, and spectroscopes, including a $1,000 Brashear spectroscope to help with Hale's senior thesis at MIT.[92] After his graduation, rather than take up one of the faculty positions that he had been offered, Hale returned home, where he convinced his father to pay for the construction and outfitting of a private research observatory for him—a $25,000 endeavor.[93] The observatory, built in 1888 next to the family home in Kenwood, had its own dedication ceremony, with noted solar astronomer Professor Charles Augustus Young being invited by Hale's father to travel to Chicago from Princeton to give the dedicatory address. The Kenwood Physical Observatory was equipped with the most advanced instruments for solar observation, Hale's principal interest at the time, and it was here that Hale undertook the research on the emission lines in the spectra of nebulae, which led to his first published paper at the age of twenty-three. William Hale would continue to play a significant role in supporting his son's efforts, including personally paying the salary of George Ritchey, the innovative telescope designer and assistant who worked with Hale, and even buying the mirror blank for what would become the sixty-inch reflecting telescope on Mount Wilson before any other support had been secured—an additional expenditure of some $10,000.[94] The use of family resources underlines the self-initiated nature of Hale's endeavors. He did not rely on any external demand for an astronomical observatory to initiate his plans; instead, he made a decision to pursue projects of his own interest and then went about finding the resources to achieve them.

Those resources would come, in one form or another, from single individuals, the first of whom, Charles Tyson Yerkes, would provide them based on a blatant and singular interest in the signaling attributes of large observatories. Yerkes was the despised robber baron behind Chicago's street railway system: he had once frankly stated that "the secret of success in my business is to buy old junk, fix it up a little, and unload it upon other fellows."[95] Needless to say, this approach to business left Yerkes with a public relations problem and a resultant interest in finding some philanthropic polish to apply to his image. When Hale, now a young faculty member at the University of Chicago, sent him a letter outlining the possibility of securing the largest refracting telescope in the world for an observatory that would henceforth bear his name, Yerkes saw an opportunity to improve his reputation and gain entrée to the circles of the Chicago elite that shunned him. Hale had come to his university perch through an arrangement that saw the Kenwood Observatory become part of the University of Chicago in return for a faculty position for Hale and a promise from the president of the university, William Rainey Harper, to secure over $250,000 for a larger observatory within two years. Within months of Hale joining the faculty, the latter part of that arrangement would be spectacularly fulfilled, thanks to Yerkes's desire for a monument and the emergence on the market of the largest refractor lens in the world.

The forty-inch lens that would become the centerpiece of the Yerkes Observatory had been born out of the enthusiasm that accompanied the height of the 1880s Los Angeles real estate boom. Edward F. Spence, president of First National Bank and president of the Board of Trustees of the University of Southern California, set out to acquire for his university a telescope that would overshadow the University of California's thirty-six-inch Lick refractor. In 1887, Spence contributed $50,000 of his own money to pursue this goal, and there were numerous fund-raising rallies and banquets held to generate interest and support.[96] A contract for a forty-inch refractor was signed with Alvan Clark & Sons, but the death of Spence and the bursting of the Los Angeles real estate bubble in 1892 precipitated a default on the agreement by the trustees, leaving the Clarks with a mostly complete, largely unpaid for set of enormous lenses. While attending a meeting of the American Association for the Advancement of Science in Rochester, Hale heard Alvan Clark tell the story of the homeless, now completed refractor and the opportunity for discovery that awaited those who might be willing to buy it from him. The young Hale was entranced with the idea and returned home to Chicago in search of a patron who could put him in possession of the largest telescope in the world. The story of how the twenty-four-year-old Hale managed to achieve this is perhaps best left to Hale himself, who described the events in his note on "The Beginnings of the Yerkes Observatory," written in 1922:

> After consulting my father, whose interest in my project was very keen, I visited several men who might conceivably be willing to provide for the telescope. But no one had the money to spare. A few days later I made another fruitless round of visits in the city. At noon, somewhat discouraged, I called at the Corn Exchange Bank to see Mr. Charles Hutchinson, then, as now, the enthusiastic friend and supporter of every such effort. After explaining my object, I asked for suggestions. "Why don't you try Mr. Yerkes?" he replied. "He has talked of the possibility of making some gift to the University, and might be attracted by this scheme." So I went at once to President Harper. . . . After a few questions he heartily approved of the attempt, asked me to write out a statement of the plan, and sent it to Mr. Yerkes. A reply came back asking us to call on him. We did so, and before the interview was over Mr. Yerkes asked us to telegraph for Clark, with whom he made a contract for the 40-inch objective. I remember with pleasure Dr. Harper's enthusiasm as we left the office. "I'd like to go on top of a hill and yell!"[97]

Although Yerkes was initially enthusiastic about the project, his keenness was short-lived. Yerkes had hoped that the telescope project would signal to Chicago society his beneficence, and he basked in the bonhomie and goodwill that followed the announcement of his planned donation. Encouraged by the public reaction, he let the cost estimate of the endeavor creep upward to $500,000 from the original $300,000 estimate made by Alvan Clark, and he was even quoted in the *Chicago Tribune* as having promised $1 million. Only one month after the announcement, however, when Hale submitted his detailed analysis of the cost of constructing and outfitting the observatory at $285,375, Yerkes refused to pay this amount.[98] His interest—like many observatory patrons before— was not in outfitting an observatory for research but simply in attaching his name to the biggest telescope in the world. He was unwilling to set up an endowment for the observatory and at the same time refused to allow other philanthropists to create endowments or donate instruments that would carry their names to "his" observatory. Although he ultimately expended some $349,000 to complete the observatory, this left it with no endowment for faculty or equipment.[99] Yerkes was evidently only prepared to pay the bare minimum for the "world's largest telescope" and had been wholly driven by its monument value.

The observatory did, however, boost his reputation, at least initially. On the dedication day in 1897, nearly eight hundred dignitaries attended the dedication ceremony of the observatory, where Yerkes—seated on a dais within its dome—reportedly blushed at the standing ovations. *The Chicago Times-Herald* suspended its regular attacks on Yerkes, saying, "Whatever opinion we may hold of Mr. Yerkes in his relations with the people of Chicago, there can be only one opinion, and that extremely complimentary, of Mr. Yerkes as the founder of the Observatory at Lake Geneva."[100] Shortly after the dedication, however, the attacks resumed, and Yerkes was ultimately driven out of town in 1900. He stopped giving any funds to the observatory in 1903, and when he died, two years later, his will allotted only $100,000 of his approximately $5-million estate to an endowment for the observatory.[101] The story of the Yerkes Observatory is an interesting counterexample to his own statement made at the dedication ceremony: "One reason why the science of astronomy has not more helpers, is on account of its being entirely uncommercial. There is nothing of moneyed value to be gained by the devotee of astronomy; there is nothing he can sell."[102] Despite Yerkes's claim, it is clear that Hale was able to sell him an observatory for the purpose of burnishing his reputation, a purpose that, temporarily at least, the observatory was able to fulfill.

Even before the Yerkes Observatory was finished, however, Hale was already thinking of building his next, even larger telescope. Within two months of securing Yerkes's promise for the forty-inch refractor and attendant observatory, Hale had become determined that what he actually needed for his research was a sixty-inch reflecting telescope.[103] For Hale, a student of spectroscopy, reflecting telescopes were particularly valuable, as they eliminated the problem of the chromatic dispersion of light that bedeviled the use of spectrometers with refractors. He often discussed the potential of building large reflecting telescopes with George Ritchey, a thirty-year-old craftsman in Chicago who was a devoted believer in reflectors and in his own ability to build them larger than ever before. Ritchey was already in the process of manufacturing a twenty-four-inch reflector, which would ultimately be housed at the Yerkes Observatory. Like Hale, however, he already had grander ambitions. By 1895, their ambitions conjoined, as Hale convinced his father to personally contract with Ritchey for the purchase and polishing of a sixty-inch mirror that would form the heart of Hale's next telescope.

Although Hale had originally intended the sixty-inch reflector to become part of the Yerkes Observatory, the death of his father in 1898, the strained relationship with Yerkes, and the poor weather at Lake Geneva meant that Hale now looked to find a new home and a new funding source for the telescope in the clear air of the mountains of California. However, his initial appeals to potential donors—like grain market mogul Norman Ream and the widow of the founder of the Union Stock Yards, Timothy Blackstone—were unsuccessful. Although the source of financial support for what would become the Mount Wilson Observatory was ultimately a single, wealthy individual, it was accessed through a relatively new type of institution that would shape the way astronomy and observatories were funded throughout the twentieth century—the large American philanthropic foundation.

In 1902, the two institutions that would finance most of Hale's astronomical ambitions in California were founded—the Carnegie Institution of Washington and John D. Rockefeller's General Education Board. There had been privately financed philanthropic organizations in America before, but the ambitions and wealth of Andrew Carnegie and John D. Rockefeller transformed the nature of the philanthropic foundation. Unlike the Smithson bequest, the funds would not be at the disposal of government employees but in the hands of the private citizens who were the foundations' board members and officers. Unlike earlier private foundations, such as the Peabody Institute in Baltimore, these new organizations were established from the start to be permanent institutions that would use the annual interest from large financial endowments to fund philanthropic causes in perpetuity. The rise of large philanthropic foundations would change the funding dynamics for American astronomy, and no one was a more significant part of that change than Hale.

The Carnegie Institution of Washington—founded by America's premier steel tycoon, Andrew Carnegie, for the purpose of supporting scientific research—would be the first foundation that Hale would cultivate in order to fulfill his ambitions. The thirty-four-year-old Hale immediately understood the new organization's inherent potential to support his sixty-inch reflector project when he read of its establishment and its unprecedented $10-million endowment ($2.1 billion in 2015 PWC-ratio terms; $7.4 billion in 2015 GDP-ratio terms).[104] He mailed the recently established Carnegie Institution Executive Committee his five-page summary of his intended project, "A Great Reflecting Telescope," complete with spectacular images of nebulae that had been taken by Ritchey on the twenty-four-inch reflector. Although there was no immediate direct response to his proposal, Hale was appointed one of the initial members when the Carnegie Institution established its Advisory Committee on Astronomy. From this position, the energetic young Hale used his influence and connections to move the foundation ever closer toward supporting his plan. Hale so deftly embedded his ambition into the culture of the organization that, on the face of it, he did not even officially request the funding for the project—he was simply responding to a request for information from Charles Walcott, secretary of the advisory committee, as to the capabilities and cost of a hypothetical sixty-inch telescope.[105]

In anticipation of the Carnegie Institution's support for his project, Hale again demonstrated his personal initiative by moving out to Pasadena in 1904 and beginning the establishment of the observatory without any official word of financial support. Although he had received a couple of small grants totaling $13,000 from the institution, this was insufficient to cover the cost of moving his team, including Ritchey, to California and setting up shop. In moving his operations to California, Hale incurred a substantial personal debt of some $27,000.[106] Hale had made this move in the expectation that it would all but force the Carnegie Institution to support his full plan to install the sixty-inch reflector—a plan that he estimated would cost $325,000 over five years. In 1905, within a year of his move, his gambit had succeeded and the Mount Wilson Solar Observatory became the first astronomical observatory to be established by a major philanthropic organization.

The endowing individual still had enormous influence in the new system, and Hale's success within it was in no small part due to Carnegie's personal interest in astronomy—an interest that Hale made sure to encourage and cultivate. Carnegie had come to his own appreciation of astronomy at least as early as 1900, when his famous "The Gospel of Wealth" singled out astronomy as one of the "Best Fields of Philanthropy" and noted the recent works of Lick and Thaw as being especially worthy of commendation:

> If any millionaire be interested in the ennobling study of astronomy,—and there should be and would be if they but gave the subject the slightest attention,—here is an example which could be well followed, for the progress made in astronomical instruments and appliances is so great and continuous that every few years a new telescope might be judiciously given to one of the observatories upon this continent, the last being always the largest and the best, and certain to carry further and further the knowledge of the universe and of our relation to it here upon the earth.[107]

Corresponding to its founder's views, the Carnegie Institution had allocated to astronomy $21,000 in its first awarding of grants, more money than to any other department of science.[108] Hale further cultivated Carnegie's interest in astronomy, sending him high-quality transparencies of Ritchey's nebulae photographs and corresponding with him on scientific subjects. The result was an enthusiastic supporter who strongly influenced his foundation's decision to support the growth and development of Hale's ambitions at Mount Wilson.

Although the Carnegie Institution would provide the majority of the funding required for Hale's third large telescope, the one-hundred-inch reflector, it would be the personal support of two wealthy individuals that would create the conditions for that support. Within a year of receiving the Carnegie Institution's support for the sixty-inch reflector, and before it was mounted in its observatory, Hale was already thinking about his next big telescope. Hale and Ritchey had developed a relationship with John. D. Hooker—a wealthy hardware-supply-company owner, oil-well investor, and an amateur astronomer in Los Angeles—and they convinced

him to fund the development of their new scheme for a hundred-inch reflector. As with the sixty-inch, Hale sought first to secure the mirror, and
the prestige-conscious Hooker agreed to provide for it with $45,000, to
be paid in $5,000 increments starting in 1906, provided that it bore his
name.[109] It was Carnegie himself, however, who enabled the project to
reach fruition. In 1910, at the age of seventy-five, the retired steel baron
personally visited the observatory and came away with a renewed enthusiasm for astronomy. On Carnegie's return to New York, he gave the Carnegie Institution another $10 million and made clear his desire that the
funds be used to support the construction of an observatory for the
hundred-inch telescope.[110] The new project would cost the Carnegie Institution over $570,000, bringing the Institution's total capital expendi-
ture on Mount Wilson, not including salaries, to over $1.4 million.[111]
While it was ultimately the trustees of the Carnegie Institution that authorized the project, it was Carnegie's influence that ensured the trustees' approval, and it was Hooker's willingness to fund the hundred-inch
mirror that helped encourage Carnegie. Even in the era of the foundation observatory, individual wealthy philanthropists often still played a
deciding role.

Hale's correspondence with Carnegie reveals how he developed the relationships with his patrons as friends and colleagues in research. His letters back and forth to Carnegie are filled with invitations from Hale to visit
Mount Wilson, invitations from Carnegie to visit Skibo Castle in Scot-
land, and books exchanged between the both of them—in one instance,
a book about cooperation on solar research from Hale and a book on the
life of James Watt from Carnegie in return.[112] Gift giving was an important part of Hale's relationship building, including presenting Carnegie
with the honor of having an asteroid named after him—671 Carnegia.[113]
Hale picked up on Carnegie's appreciation of astronomical lantern slides
early and repeatedly sent Carnegie cabinets with new slides that included
not only astronomical photographs from the observatory but also attractive photographs of the stunningly situated Mount Wilson Observatory.[114] In addition to the prestige conferred on Carnegie, these would
become the principal physical currency of the relationship, as one of Carnegie's glowing letters indicates:

Dear Friend,

We hear of some wonderful new slides since you sent us the tresured [*sic*] case with what you had then. We pray for the others—do send them to us. These have been enjoyed by many distinguisht [*sic*] people. Lloyd George was held in a trance one nite, and after the display closed and the lites [*sic*] were turned up he still in a trance as it were slowly said "Never in all my life have I been so entranced." We owe you much.

Kindest regard to wife and dauter [*sic*], Very truly yours, Andrew[115]

Although their relationship soured when Hale pushed hard on Carnegie to fund a building for the National Academy of Sciences, Hale had fastidiously developed a friendly and collegial relationship with his first philanthropic patron, and it paid off handsomely.

Hale had also maintained a strong relationship with Carnegie's wife, Louise, including after the death of her husband—giving her gifts of books and noting in one of his letters to her that "we at Mount Wilson appreciate increasingly the unequalled opportunity [Mr. Carnegie] gave us to extend our research into the depths of space."[116] Although Mrs. Carnegie was an old family friend by this time, an undated newspaper clipping that Hale kept suggests there may have been other considerations:

LAW REDUCES RESIDUARY BEQUEST TO CARNEGIE CORPORATION TO LESS THAN $11,000,000—WIDOW'S SHARE $11,338,847[117]

Ever the careful strategic planner, Hale's incredible correspondence regime—one diary entry from 1922 shows a week with twenty letters written every day—and the relationships he built and maintained through it, was one of the critical success factors in his astronomical entrepreneurship.

Through his correspondence, Hale showed a canny capacity to cultivate the personal astronomical interests of individual philanthropists. This did not, however, necessarily mean a cultivation of scientific interest in astronomy. The description of Mount Wilson in Carnegie's autobiography displays an amateur's exuberance for the subject but also a somewhat limited understanding of the state of contemporary astronomy, which Hale had done little to correct:

> There is but one Mount Wilson. From a depth seventy-two feet down in the earth photographs have been taken of new stars. On the first of these plates many new worlds—I believe sixteen—were discovered. On the second I think it was sixty new worlds which had come into our ken, and on the third plate there were estimated to be more than a hundred—several of them said to be twenty times the size of our sun. Some of them were so distant as to require eight years for their light to reach us, which inclines us to bow our heads whispering to ourselves, "All we know is as nothing to the unknown." When the monster new glass, three times larger than any existing, is in operation, what revelations are to come! I am assured if a race inhabits the moon they will be clearly seen.[118]

Although Carnegie's endowment of the Carnegie Institution of Washington had undoubtedly been in part motivated by the prestige and legacy concerns of the unpopular steel magnate, his support for astronomy seems to have been driven by genuine philanthropy and intrinsic motivations: it is telling that neither Carnegie nor his foundation ever asked that the Carnegie name be attached to the observatory or telescope.[119] It was Carnegie's personal stamp of approval that drove the funding of the hundred-inch telescope, even if it was the trustees of his foundation that executed the project. The source of support for Hale's final and largest telescope, the two-hundred-inch telescope at Palomar Mountain, would also be a philanthropic foundation—this time, however, the motivating force would come not from the founder, but directly from an individual within the official foundation bureaucracy.

The two-hundred-inch telescope makes a fitting end point in this study of U.S. observatories—not only because it is the last large optical telescope to be built in America before the Cold War but also because it marks a turning point in the development of professional science bureaucracies. The role of the patron was beginning to fade into the background, eclipsed by that of professional decision-makers working independently with significant discretion. In the case of the two-hundred-inch telescope, although it was John D. Rockefeller who established the Rockefeller Foundation funding body, he played, at most, a peripheral role in approving the project. The impetus and support within the foundation came instead from Wickliffe Rose, the president of the foundation's General Education Board, a man who had no technical training but nonetheless saw himself as a new type of professional—a science administrator. In 1923, Rose convinced the Rockefeller Foundation trustees to provide $28 million for an International Education Board that would support the development of science.[120] In contrast to Carnegie's more intimate involvement with his foundation's philanthropy, Rockefeller took a largely hands-off approach that gave Rose, as the head of the new board, almost complete control over the disbursal of the funds. This delegation of authority to the professional science manager would emerge again in the more circumscribed, but nonetheless meaningful, control that senior managers at NASA and other government science agencies would have to pursue and fund projects with significant personal discretion.

Although the control of resources for astronomy was shifting to professional research managers like Wickliffe Rose, the challenge for a project entrepreneur like Hale was still largely the same—to convince an individual or a small group of individuals to provide the financing for an ambitious project. After the success of the hundred-inch Hooker reflector, Hale's thoughts had turned to even larger mirrors and, along with Mount Wilson astronomer Francis Pease, he began to sketch out a design for a mammoth three-hundred-inch reflector. When H. J. Thorkelson of the General Education Board of the Rockefeller Foundation visited Pasadena in October 1926, Hale made sure that he received a tour of the Mount Wilson facilities and was shown the designs for the new telescope, knowing full well that Thorkelson would mention the plans in his report back to the foundation. When the editor of *Harper's Magazine* asked him for an article on an astronomical topic, Hale was ready. His article, "The Possibilities of Large Telescopes," published in early 1928, was part romance to the heavens, part primer on large reflecting telescopes, and all part of the pitch for his next observatory. He had a preprint copy of the article sent to Rose and wrote him a long letter expanding on the topic. Hale concluded his letter by asking if the General Education Board might be able to provide a small amount of funding to study the feasibility of casting the types of large mirrors that he had outlined in his article. The result was an invitation from Rose to discuss the matter with him in New York, and within a month, Hale had met with Rose and enlisted his enthusiastic support for the construction of the new largest telescope in the world.

Although the personal correspondence and private discussions of Hale and Rose were crucial in establishing the financial support for the new telescope, it was Hale's *Harper's* article that framed the public perception of the need for a new large reflecting telescope and that, in turn, influenced the eagerness of Rose and the Rockefeller Foundation to support the initiative. The rhetoric that Hale employed in the article provides insight into the themes he thought would be effective in eliciting financial support for the project. The direct connection that Hale drew, not only between astronomical observatories and the rich and powerful, but also between astronomy and exploration, is illustrative of the grand narrative that Hale wove into his observatory fund-raising efforts. The opening lines of his article are rife with allusions to legend and legacy and references to the themes and objects of exploration:

> Like buried treasures, the outposts of the universe have beckoned to the adventurous from immemorial times. Princes and potentates, political or industrial, equally with men of science, have felt the lure of the uncharted seas of space, and through their provision of instrumental means the sphere of exploration has rapidly widened. If the cost of gathering celestial treasure exceeds that of searching for the buried chests of a Morgan or a Flint, the expectation of rich return is surely greater and the route not less attractive. Long before the advent of the telescope, pharaohs and sultans, princes and caliphs built larger and larger observatories, one of them said to be comparable in height with the vaults of Santa Sophia. In later times kings of Spain and of France, of Denmark and of England took their turn, and more recently the initiative seems to have passed chiefly to American leaders of industry. Each expedition into remoter space has made new discoveries and brought back permanent additions to our knowledge

of the heavens. The latest explorers have worked beyond the boundaries of the Milky Way in the realm of spiral "island universes," the first of which lies a million light-years from the earth while the farthest is immeasurably remote.[121]

Hale made explicit the opportunity for wealthy individuals to become one of the "adventurous" that explored "the outposts of the universe." Although he carefully situated his narrative within the context of the legacies of the rich and powerful figures of history, he deftly makes his most direct appeal to personal curiosity: "Lick, Yerkes, Hooker, and Carnegie have passed on, but the opportunity remains for some other donor to advance knowledge and to satisfy his own curiosity regarding the nature of the universe and the problems of its unexplored depths."[122] Hale makes clear the ability of observatories to create lasting legacies, but he also appeals directly to, and looks to cultivate, a personal interest and exhilaration in the wonders of the universe. He emphasized that scientists were but one element involved in the exploration of the heavens; politicians and captains of industry were also part of the fellowship and critical to its success. With the tone of the narrative set, Hale proceeds with an explanation of his accomplishments to date, the discoveries in stellar evolution, spiral nebulae, and fundamental physics that awaited larger telescopes, and detailed descriptions of the technologies and instrumentation required to properly build a two-hundred- or three-hundred-inch reflecting telescope. In a short eight-page article, in a magazine for the general reader, Hale laid out his plans for his next great observatory within a narrative of cosmic exploration. He made apparent the opportunities for legacy, knowledge, and adventure that were open to anyone with sufficient resources to partake. It was with this groundwork in place that Hale met with Rose to solicit his support for what would become the Palomar Observatory.

Through his position at the Rockefeller Foundation, Wickliffe Rose controlled the resources required to participate in the explorations that Hale had described so enticingly in his article. Hale's article had also struck the right chords in relation to Rose's desire for legacy. Hale records that Rose was immediately taken with the project, wanting nothing but the largest and most remarkable telescope, encouraging Hale to consider the three-hundred-inch option and suggesting that as much as $15 million could be spent on the project.[123] Hale and Rose soon settled on $6 million for the more conservative two-hundred-inch, although with the inevitable cost overrun the final cost was over $6.55 million.[124] In a letter to a friend, Hale described his surprise at the stunning success which his article had generated: "An article of mine on large telescopes, shot like an arrow into the blue, seems to have hit a 200" [-inch] reflector, and I have been forced to take to the air myself and try to parachute it to earth."[125] The tones of historic adventure that Hale had set in his article had clearly resonated with Rose; his letters to Hale expressed his belief that the project "was a thrilling adventure" and that it "was a privilege to live in an age where such a thing can be done."[126]

The support for the project within the Rockefeller Foundation emanated almost entirely from Rose. Rockefeller was cognizant and supportive of the project but had only a limited interest in the subject, sending Hale only a single perfunctory letter of commendation and commenting publicly: "I have no competence in the field of astronomy. Six million dollars *is* a large sum of money, but I have complete confidence in Mr. Rose and the trustees, and if after careful investigation they decided that it is the wise thing to do, there certainly will never be any criticism from me."[127] Rose's eagerness and the unprecedented large size of the grant seem to have been in part due to Rose's impending retirement and his corresponding desire to use the remaining, uncommitted funds of the International Education Board to support the type of scientific research that he most valued, anticipating that his successor might favor other areas of sponsorship.[128] That an individual administrator would be able to make such a large commitment speaks to the changing dynamics of philanthropy in the early twentieth century. As the century developed, powerful science and technology administrators—whose appointments gave them control over significant resources, albeit often only for a limited time— would play an increasingly important role in the funding of the explorations of the heavens. While the Palomar Observatory represents the increasing bureaucratization of science funding in America, it is important to recognize that the decision to fund the project was still largely the result of Hale engaging the interest and ego of one man.

The signaling value of the Palomar Observatory was also keenly appreciated, as evidenced in the near collapse of the project due to the difficulties of cooperation between the Carnegie Institution and the Rockefeller Foundation. With the notorious personal rivalries between Carnegie and Rockefeller, there was no chance that the Rockefeller Foundation would directly fund a project under the auspices of the Carnegie Institution's Mount Wilson Observatory. On the Carnegie side, the institution's president, John Campbell Merriam, objected strongly to the notion that the Rockefeller Foundation might build a new world's largest telescope with the direct assistance of the honorary director and staff of his institution's own observatory. It took famed international diplomat and Nobel Peace Prize winner Elihu Root to negotiate an arrangement whereby the two funding institutions could work together on the development of the new observatory and share in its signaling value.

The agreement put the two-hundred-inch telescope under the direct management of an institution that Hale had been instrumental in developing during his time in Pasadena—the California Institute of Technology. In 1907, Hale had joined the board of trustees of the small Throop Polytechnic Institute and begun the process of transforming it into a university that focused on scientific and technical excellence. Hale had already managed to entice famed American physicist Robert Millikan to Caltech and now had lured in the largest telescope in the world as well, provided that the institute could raise a sufficient endowment to cover the annual operating funds of the observatory. Although such an endowment was a tall order for the young university, Hale and Millikan had long cultivated a relationship with Henry M. Robinson, chairman of the First National Bank of Los Angeles and Caltech trustee. They were able to convince him to commit to provide for the estimated $1.5-million annual operating expenses.[129] It had taken Hale less than four months, from his meeting with Rose on March 14, 1928, to Robinson's announcement of support on June 11 of the same year, to secure the funding for his largest and most expensive observatory.

Hale's relationship with Robinson has been one of the least investigated in the secondary literature, despite the importance of their connection in securing the final funding for the two-hundred-inch telescope. It bears further investigation. It began with a gift-giving letter from Hale to Robinson in 1915, in which Hale offered a directorship position in the Pasadena

Music and Art Association, one of the many societies Hale was involved with.[130] The two began to correspond regularly, with Hale providing a mentorship role in scientific matters to the intellectually curious Robinson, advising him to join the American Association for the Advancement of Science and continuing to provide Robinson with prestigious gifts, such as a letter of recommendation for admission to the University Club of New York. In return, Robinson performed a rather unique service for the ever-scheming Hale: he used his authority as chairman of the First National Bank of Los Angeles to provide Hale with extensive private financial reports on high-net-wealth individuals in the Los Angeles area.[131] For over a decade, Hale cultivated his relationship with Robinson to the point where they were sufficiently close as friends and colleagues—borderline partners in crime even—so that when the time came for a major financial request for Hale's last great project, the response from Robinson was sure to be quick and positive.

Although the financing was in place, there were still decades of labor and crises to endure before the 200-inch telescope saw first light in 1948. The technical challenges that had to be surmounted were immense: the fabrication of the giant mirrors, the first space exploration "clean-room" at the grinding laboratory, the transport of the mirrors across the continent with crowds flocking to the railroad tracks to watch them pass, and the twenty-year struggle to make the observatory a reality.[132] The technological complexity and the popular response to the 200-inch reflector shows how the observatories were beginning to approach the drama of the earliest space probes, which began to fly into orbit less than ten years after the Palomar Observatory's completion. This timing highlights the continuum in the narrative of space history. There was institutional continuity as well. The California Institute of Technology would go on to develop the principal NASA center for robotic space exploration, the Jet Propulsion Laboratory, assisted in part by the expertise in space research and project management that Hale and the Palomar Observatory had nucleated. The 200-inch reflector, named the Hale Telescope in honor of the man whose vision created it but who did not live to see it in operation, would maintain its supremacy as the world's largest reflecting telescope well into the Space Age. It was not until 1976 that it was finally surpassed in size—if not necessarily in observing quality—by the 236-inch Bol'shoi Teleskop Azimultal'nyi in the Soviet Union. The Palomar Observatory is a true embodiment of economic, institutional, and motivational continuity between the history of American astronomy and the Space Age.

The Palomar Observatory also represents the end of an era, however, as it was the last major observatory to be built by private sources in America for almost half a century. With an increase in government sources of science funding during and after World War II, and with the formation of the National Science Foundation and NASA, astronomers began to rely principally on government funds, often secured through academic coalitions like the Association of Universities for Research in Astronomy (AURA).[133] Between the completion of the two-hundred-inch telescope in 1948 and completion of the first W. M. Keck Observatory in 1992, there were no new large privately financed observatories in America, although there were a number of significant ground-based and space-based observatories funded through government programs. When private financing for large astronomical observatories reemerged, it took on a familiar tone. In 1985, the W. M. Keck Foundation, established out of the wealth of Superior Oil Company founder William Myron Keck, gave $70 million, nearly a fourth of its assets, to Caltech to finance the design and construction of the world's largest telescope on top of Mauna Kea in Hawaii.[134] As in the case of the Palomar Observatory, one foundation trustee in particular had been instrumental in steering the foundation toward the telescope project—Simon Ramo, cofounder of the Ramo-Wooldridge Corporation, which had served as lead contractor on the Atlas rocket that launched John Glenn into space.[135] The model of privately funded, foundation-supported astronomy, which Hale had been so instrumental in developing at the beginning of the twentieth century, would be employed to build in America the largest telescope in the world once again.

With our review of the economic history of American astronomical observatories from the nineteenth to the mid-twentieth centuries now complete, it is worth comparing it to the history of funding for scientists and scientific research in astronomy during the same period. Although astronomy was a popular subject for private support in nineteenth-century America, it was the visible, physical infrastructure of astronomy—not the scientists—that received the vast majority of the funding. Indeed, even the emphasis in the literature on the support for the "practical aspects" of American astronomical sciences is shown to be somewhat misplaced in economic context: although there was some interest in the practical applications of astronomy, most notably from the government, the most significant early support for astronomical activities came largely from nonpractical intrinsic interests and signaling. As has been noted by John Lankford, the volatile and often piecemeal nature of research funding—practical or otherwise—meant that the growth of the astronomical community in America lagged far behind the development of its physical infrastructure for much of the nineteenth century.[136] Edward Pickering, director of the Harvard College Observatory, observed that there were far too many instances "of observatories without proper instruments, large telescopes idle for want of observers, and able astronomers unprovided with means of doing useful work."[137] Although America had 102 observatories by 1907, it did not have a single general endowment fund for astronomical or astrophysical research.[138] While the history of science literature has tended to see this as a "failure" of the funding process, as did the scientists at the time, it is possible to have another perspective on it—that the interest and popular enthusiasm that drove the early economic history of astronomy was not in astronomy as a science but in the exploration of the heavens as a symbolic and personally enjoyable activity.

The exception that proves the rule on this point is provided by the case of Catherine Wolfe Bruce. From 1889 to 1900, she gave $174,275 for astronomical and astrophysical research.[139] She received little public attention but was acclaimed privately by astronomers on her death as "one of the most sympathetic and generous patrons astronomy has ever known."[140] Indeed, she emphatically avoided the public eye, seldom leaving her house or receiving visitors, living in obscurity on her family fortune, and explicitly rejecting publicity. She gave fifty-four grants from $50 to $25,000 and even assumed the responsibility for Professor Frost's salary at the Yerkes Observatory when Yerkes himself refused. Importantly, all of her donations were made during her lifetime, with her will containing only personal bequests.[141] She subsidized publications and investigators and focused on the needs of astronomy that, unlike telescopes and observatories, would not raise public notice. She also differed from the majority of other philanthropic benefactors of astronomy in her intense personal involvement, making her a classic example of the role that intrinsic interest in astronomy played in its financing. Her very uniqueness, however, also presents important evidence for the limited interest in the funding of astronomical activities unrelated to signaling during the period.

The dominant role of signaling interests rather than scientific interests for the observatory patrons of the nineteenth century underlines the importance of the "astronomical entrepreneurs"—individuals such as Mitchel and Hale—who were able to enter into exchanges with patrons interested in monuments and the personal exploration of the heavens and, on occasion, to use those interests to secure support for their personal interest in scientific research. It also emphasizes the vital role of government support for scientific research in astronomy that ultimately emerged in the twentieth century. Although government patrons are also susceptible to prioritizing prestige over pure science, the scientific peer networks that are more habitually relied upon by government institutions as advisory boards in funding decisions make for strong advocates for funding research, scientific equipment, and the general support of scientists. Thus, while private patronage is a long-standing and robust phenomenon in the history of American astronomy, it is important to remember that it is also a phenomenon that had a tendency to privilege edifices over edification.

The economic information presented here also situates the history of American astronomy within the overall context of American space history and space exploration activities. Understanding space exploration within this wider context reorients a basic assumption of the literature. An evaluation of American space history that includes an analysis of the funding of American observatories will see a more nuanced picture—one that shows private funding to have been dominant for over a century—than an evaluation of American space history that begins its analysis of space exploration activities with Sputnik and the dominantly governmental support for space exploration in the Cold War space race. Motivating forces similar to those behind the funding of astronomical observatories—intrinsic interest and signaling—would propel space exploration into the Space Age. It is the demonstration of a comparable scale in expenditures between astronomical observatories and modern space exploration projects, however, that gives real economic meaning to the suggestion of a continuum in space history between telescopic and rocket-based space exploration. That American observatories in the nineteenth century were as resource-intensive, relative to the scale of the economy at the time, as even large American space missions in the twentieth century is an important fact that has received little attention to date. It is also true, however, that even the most expensive projects of the earlier era—the Lick and Palomar Observatories—had expenditures that were two orders of magnitude smaller than those of the Apollo program. The continuum of space expenditure thus still shows a significant step change in the mid-twentieth century due to the signaling context being elevated from the level of individuals and urban communities to the level of competition between superpower nation-states. Nonetheless, following the threads of space exploration back to the origins of American astronomical observatories provides a much richer historical context for understanding the development of twentieth- and twenty-first-century space exploration, as well as a number of instructive precedents.

The threads of this Long Space Age perspective are many. The threads are individuals and institutions: the astronomer Samuel Pierpont Langley, director of the Allegheny Observatory, who became a crucial figure in early American aeronautics and is the namesake of NASA's Langley Research Center; government astronomer Simon Newcomb, who organized the 1872 American transit of Venus expeditions and wrote spaceflight fiction; the Smithsonian, born from John Quincy Adams's passion for astronomy, which would be the first institution to provide funding for Robert Goddard's work in liquid-fuel rocketry; George Ellery Hale, inspired by Verne's tale of space travel, who not only built his era's most powerful instruments of space exploration but who supported Goddard's work and shaped the small Throop Polytechnic Institute into Caltech, from which emerged NASA's Jet Propulsion Laboratory. The threads are also the motivating trends: of nations, communities, and individuals looking to signal their development through space exploration; of the wealth of self-made industrialists, whether from the railroads or the Internet, being dedicated to establishing organizations that would contribute to the exploration of the heavens; of individual astronomers and engineers dedicating their lives and intellects to the mobilization of the resources required to achieve their visions. The threads are also economic: the comparable economic significance of nineteenth-century observatories and modern spacecraft, and the progressively increasing complexity of space exploration projects. These threads must be woven together and the history of American space exploration examined in its entirety if we are to fully appreciate the long-run forces propelling our voyage out into the solar system. In the long historical perspective, the trend in the late twentieth and early twenty-first centuries toward increased funding for space exploration projects coming from the private sector—specifically from wealthy individuals such as Paul Allen, Jeff Bezos, and Elon Musk—is understood not as a new emerging phenomenon but rather as the reemergence of a dominant thread in space exploration that dates back to over a hundred years before Sputnik. Incorporating the history of astronomical observatories into the overall narrative of American space history shows that, in fact, it has been private sources that have supplied the resources for the nation's exploration of the solar system and the universe for most of its history to date.

3

SPACEFLIGHT, MILLIONAIRES, AND
NATIONAL DEFENSE: ROBERT GODDARD'S
FUND-RAISING PROGRAM

The salvation of the planet, as everybody was now convinced, depended
upon the successful negotiation of a gigantic war fund, in comparison
with which all the expenditures in all of the wars that had been waged
by the nations for 2,000 years would be insignificant.

—*Garrett P. Serviss,* Edison's Conquest of Mars, *1898*

"In the history of rocketry, Dr. Robert H. Goddard has no peers.
He was first. He was ahead of everyone in the design, construction, and
launching of liquid-fuel rockets which eventually paved the way into space."[1]
So testified Wernher von Braun to the Senate Committee on Aeronautical and Space Sciences in 1970. Although modern historiography has
positioned Goddard as one of the three "fathers" of spaceflight—along
with the German-Romanian Hermann Oberth and the Russian Konstantin Tsiolkovsky—his contribution to liquid-fuel rocketry development
was of a different order of significance. While the contributions of Oberth
and Tsiolkovsky were largely theoretical and cultural, Goddard was the
first to achieve flight with a liquid-fuel rocket, as well as the first to develop a spaceflight technology development program worthy of the name.
Goddard was also ahead in another important aspect that is not well
recognized: he was the first to acquire and apply significant financial resources to the problem of spaceflight.

Goddard has hardly suffered from a lack of historical investigation.
A number of expert essays and books on Goddard have been written,

including the notable biographies by Milton Lehman and David Clary.[2] Here we will build on these works and focus on his career from an economic perspective. Goddard was internally driven from an early age to dedicate his effort and intellect to the problem of spaceflight. He was also fortunate in having found others—most notably Charles Lindbergh and Harry Guggenheim—who shared these motivations and supported his projects out of personal interest. Historians have focused on the importance of these benefactor relationships in painting a picture of Goddard as the consummate lone inventor, dedicated to spaceflight, unwilling to compromise on his projects, and reliant on those generous private patrons for support.

While Goddard was indeed dedicated to spaceflight and relied on private patronage for much of his career, he was in fact quite entrepreneurial and pursued one source of support above all others—military funding. Goddard's first approach to the Smithsonian was prompted by his bid for support from the navy in the First World War and, during that war, he was the first to use an expansion in armaments expenditure to fund spaceflight work. He also continued to pursue military contracts between the wars and made the choice to leave the comfort of his Guggenheim-funded project to pursue instead large military contracts in the Second World War. This chapter compiles a more complete record of Goddard's funding history than has previously been available, illustrating that his total military funding, which he received for only a few years of his career, rivaled the entirety of his funding from other private and public sources. In short, it argues that Goddard's "spaceflight fatherhood" peers could more properly be considered to be Wernher von Braun in Germany and Sergei Korolev in Russia—not only because they all led their nations in the early practical development of liquid-fuel rocketry but also because they all saw an exchange of their services with the military for financial support as the quickest path to the stars.

At the age of sixteen, in 1898, Robert Goddard read the *Boston Post*'s serialized version of H. G. Wells's *The War of the Worlds,* in which the events had been transposed to Boston from London. Wells's realism in describing an advanced space-faring race from Mars made a deep impression on Goddard. Within a year of reading Wells's story, on October 19, 1898, while lying in the branches of a small cherry tree on his family's property, Goddard had a vision of a spacecraft flying to Mars, which resulted in a lifelong commitment to the development of a machine that could make that vision a reality.[3] So goes the traditional narrative of Goddard's decision to pursue spaceflight. Goddard was also inspired, however, by another work of fiction, one that has received almost no attention in the literature and that would have presented a striking picture of the economics of spaceflight to the young space pioneer.

Garrett P. Serviss's *Edison's Conquest of Mars* was a sequel to *The War of the Worlds* and was immediately commissioned by the *Boston Post* and serialized following the popular success of Wells's story.[4] Goddard read both of them in succession and, in his "Material for an Autobiography" written in July 1927 and 1933, referred to the influence that the two works had on him, writing that, like *The War of the Worlds*, "Garrett P. Serviss's *Edison's Conquest of Mars*, gripped my imagination tremendously."[5] The novel begins in the period right after Wells's Martian invasion has been defeated and tells the story of the Earth's counterattack on the Red Planet. The heroes of the story are "a few dauntless men of science," including Lord Kelvin and Wilhelm Röntgen, and led by Thomas Edison. Together they invent an electrical antigravity device and use it to develop a spaceship and a handheld disintegrator in order to repel the expected second wave of Martian invaders. A massive fleet of spaceships is built and sent off to Mars—with encounters on the Moon and a Martian asteroid mining colony along the way—where through many battles and thousands of deaths on all sides, the Earth fleet breaches the walls of the famous Martian canals and causes a deluge that lays waste to the Martian civilization. Goddard's boyhood appreciation for a pulp-fiction tale of interplanetary conquest does not mean that he himself must have harbored such martial ambitions. Nonetheless, while Oberth and Tsiolkovsky both traced their inspirations back to the peaceable tale of Jules Verne's *From the Earth to the Moon*, Goddard traced his back to two stories that integrally linked spaceflight and war—and was, coincidently or not, the only one of the three to have entered into numerous alliances with military patrons to advance his work.

Edison's Conquest of Mars also included an extensive description of the fund-raising process for the development of Edison's space fleet, which may have influenced Goddard's thinking. *From the Earth to the Moon* also included details of how Verne's protagonists in the "Baltimore Gun Club" raised the funds required for their massive cannon to launch a projectile at the Moon, but the two descriptions have a significant difference. While the Baltimore Gun Club is described as succeeding through an appeal to intrinsic interest and global goodwill—receiving its funding from individual volunteer contributions from around the world in a manner similar to some of the civic observatories—it is an appeal to the imperative of military defense that raises the funds for Edison. Just as significantly, the funds raised are vastly larger—"twenty-five thousand millions" of dollars.[6] For perspective, $25 billion dollars in 1898 was more than twice the national GDP of the United States at the time.[7] The United States was the largest contributor to the Martian conquest fleet, providing $2 billion of the total amount, edging out the United Kingdom at $1.5 billion and Germany at $1 billion. With a hundred ships in the Edison fleet, the $250-million cost of each vehicle was roughly equivalent to the $375-million total cost of the Panama Canal, the single most expensive construction project in United States history when completed in 1914.[8] The economic message of *Edison's Conquest of Mars* was clear: if spaceflight is considered critical to the defense of the nation and the planet, then there is almost no limit to the expenditure on spaceflight that might result. As the precocious young Goddard sat in his cherry tree and imagined the launch of his Mars-bound vehicle less than a year after reading Serviss's work, it is easy to imagine him thinking about this message as he contemplated obtaining the resources to achieve his dream.

In 1898, however, Goddard's relationship with the military was still many years off. Although Goddard would eventually draw more funding than any contemporary American researcher in the field of liquid-fuel rocketry, the resources he employed to develop his initial ideas and designs for spaceflight systems were limited to his own efforts, his teaching salary, and the support of his family. Born into a middle-class family in Massachusetts, Goddard, although delayed in his schooling due to concerns over his health, had a supportive environment for his inventive inclinations. His father was an inventor of small machines and supplied enthusiasm and support for Goddard's technical curiosity, including the provision of a telescope, a microscope, and a subscription to *Scientific American*.[9] Goddard also demonstrated an early entrepreneurial streak, coming up with business schemes in his early teenage years for a new type of soap, artificial diamond manufacturing, and a large-scale frog hatchery. He would ask his father for money to fund these ventures and would write business letters to his tolerant father addressed to "Gentlemen of the Company" and signed "The Manager," until he graduated from high school in his early twenties. It was in this curious family business context that Goddard, after his cherry-tree vision, began to keep a folder entitled "Aerial Navigation Department" that would come to contain many of his ideas for space travel.[10]

Goddard began to work seriously on the problem of spaceflight while pursuing his undergraduate and graduate degrees at the Worcester Polytechnic Institute and Clark University, and as a postdoctoral fellow at Princeton University. He worked through the problem and compiled his thoughts in a series of five notebooks from 1906 to 1915. In these notebooks, each with high-minded headings such as "Navigability of Interplanetary Space" and "Undertaken to find a way to learn in detail of the physical characteristics of the neighboring planets," Goddard worked out the suite of technologies he believed would be required to enable spaceflight and space habitation. One of his early concepts included a series of nested guns firing toward the Earth and propelling the system upward into space, a concept that would lead him to his initial failed attempts at creating a cartridge-based solid-fuel rocket capable of attaining high altitudes. Limited at this point only by the laws of physics and his imagination, his wide-ranging consideration of the problem led him to investigate a multitude of technologies that he believed would be required for space travel. While still a student, he considered concepts for suspended animation, the use of solar energy for in-space transportation, ion propulsion, the production of hydrogen and oxygen for fuel on the Moon, and, of course, liquid-fuel rocketry. He sketched out numerous system designs and performed the initial calculations that provided him with his initial convictions on the most promising paths of technology development. It was in this period, prior to any financing or external resources, that Goddard developed the majority of his ideas and his approach to the problem of spaceflight.

It was also in this period that Goddard articulated his philosophical motivation for his life's effort to set a course for the stars. In an outline for an article entitled "The Navigation of Interplanetary Space," written in 1913 at the age of thirty-one while recovering from his first fight with tuberculosis, Goddard described what he considered to be the "economic" argument for spaceflight: "From an economic point of view, the navigation of interplanetary space must be effected to ensure the continuance of the race; and if we feel that evolution has, through the ages, reached its highest point in man, the continuance of life and progress must be the highest end and aim of humanity, and its cessation the greatest possible calamity."[11] To achieve this end he believed humanity would learn to move throughout the solar system, launching from planet to planet, using the hydrogen and oxygen of these planets for fuel and using their metals to build additional devices for travel and habitation. In an unpublished work entitled "The Last Migration," Goddard described million-year journeys to other star systems with the aid of energy from "atomic disintegration"—the human passengers being transported in stasis, as successive generations or as a granular "protoplasm" of "such a nature as to produce human beings, in time, by evolution."[12] His most striking description of near-term space development is contained in a confidential report to the Smithsonian in 1920, although the concept is discussed in numerous places in his earlier notebooks: "The best location on the moon would be at the north or south pole, with the liquefier in a crater from which the water of crystallization may not have evaporated, and with the power plant on a summit constantly exposed to the sun. Adequate protection should be made against meteors by covering the essential parts of the apparatus with rock."[13] This description of a lunar base—at a polar crater where near-perpetual sunlight can be used for energy and where the cold traps of a permanently shadowed crater have allowed for significant reserves of water-ice—is almost identical to NASA lunar-base concepts almost a century later.[14] Prior to conducting any of his rocketry experiments or receiving any funding, Goddard had already developed a vision for a future in space that was as expansive and detailed as that of any of his predecessors—not to mention many of his successors. It was in a conscious attempt to help create that future that Goddard began his first experiments with rockets, drawing first from his own finances and the resources he had at his disposal.

The first external resources that Goddard had for experimentation and development were provided through his position as a professor of physics at Clark University. He had begun his conceptual development of the liquid-fuel rocket in 1909 while at Clark for his doctorate and included the concept in a patent in 1914, when he became an assistant professor at the university. His first experiments, however, were related to his work on a multiple-charge solid-fuel rocket that he believed could reach orbit.[15] In 1914, he began to use his own funds—derived from a salary of $1,125 as an assistant professor and $1,500 the following year when he was appointed head of the department—in order to amass a collection of solid-fuel rockets and to initiate static tests to determine the relative efficiency of different fuels and designs.[16] During this time, he performed some fifty rocket tests, using Clark's shop facilities. Using only his salary, he was limited in the work that he could do and relied on graduate student assistants for help, as well as favors from local mechanics and industrial labs. Graduate student labor remained a nontrivial component of his investigations even when he began receiving formal support for his research from external organizations. At least a dozen Clark graduate student theses were related to Goddard's research on interplanetary flight—including early experimentation with electrical ion-propulsion techniques—although he never revealed to them that spaceflight was the purpose of their projects.[17] It was thus his own modest professorial salary, the university's facilities, and its pool of graduate student labor that enabled Goddard to conduct the first experiments. These experiments confirmed his faith in the potential to develop multiple-charge solid-fuel rockets of greater efficiencies that could reach orbital velocities.

As Goddard began to rack up experimentation costs, he also began to consider alternative sources of funding that might help him develop his rocket. His first thought was the military. As early as July 25, 1914, he tried to interest the navy in his multiple-charge solid-fuel rocket, writing to Josephus Daniels, secretary of the navy, and offering to develop his rocket as a sort of aerial minelayer that would leave continuous flak along its trajectory with each successive charge.[18] Daniels and later also the acting secretary of the navy, Franklin D. Roosevelt, responded with interest, especially with regard to the method of wireless control that Goddard indicated could be added to the device. They both invited him to demonstrate his device. But, given that it was not yet developed to any level of demonstration, he let the correspondence expire. He nevertheless kept his dream of military funding very much alive. On October 19, 1915, the "Anniversary Day" of his cherry-tree epiphany, he penned a pronouncement on how he intended to approach the financing of his project: "Have the Navy or Army Department develop it for coast defense etc. provided they will allow research at high altitudes etc. with it, under government control and assisted by foundations, if necessary."[19] Within a year of writing this note and with the help of his department head, he would begin to approach the military in the hope that the increase in hostilities between America and Germany would give him the opportunity to further his spaceflight program.

The standard narrative of Goddard's emergence from academia into the world of external funding begins with his submission, unsolicited, of the first funding proposal for spaceflight technology development to the Smithsonian Institution in 1916. In fact, however, Goddard's initial letter to the Smithsonian was motivated by his pursuit of military funding, which was already under way. Arthur Webster, the head of the Clark University Physics Department, was a member of the Naval Consulting Board and a close confidant of Goddard. Although Webster had initially been skeptical of Goddard's rocketry work, Goddard's enthusiasm and experimental results convinced him. He was soon eager to present Goddard's cartridge-based, high-altitude solid-fuel-rocket proposal to the board and began to make the necessary preparations. Goddard's first letter to the Smithsonian, sent on September 27, 1916, four months before the official break in U.S.-German relations, was prompted by anticipation of this development and therefore placed the potential military applications of his rocket design at the forefront: "This communication I had intended sending a little later, but I feel that it would not be desirable to delay any longer. Incidentally, I think it would be best not to make it public. . . . My reason for writing just now is the following; My device will be capable of propelling masses, such as explosives, for very great distances, and hence would very likely be useful in warfare."[20] Goddard was not, however, contacting the Smithsonian directly in a military context, although he likely had in mind that the submission might assist him in gaining military support. Rather, his sales pitch to the Smithsonian emphasized the scientific applications of the technology: "My other reason is that the device will, I am certain, be of very great importance to pure science, especially to meteorology. . . . In short, the exclusive use of the device for warfare would, I am certain, be a loss to science."[21]

With the enthusiasm of Webster seeming to put in motion matters on the military side, Goddard was approaching the Smithsonian so the two sources of support that he envisioned in his 1915 note would be pursued in parallel. He thus asked the Smithsonian if it would consider organizing a committee, either independently or in collaboration with the Naval Board, to evaluate his concept, and, should the committee report favorably, would the Smithsonian "take upon itself the recommending of a fund, sufficiently large to continue the work, either from a society such as the National Geographic Society or from private individuals?"[22] He made no mention of spaceflight, although he did indicate that it should be possible to reach an altitude of 232 miles—well above the atmosphere of the Earth.[23] Goddard evidently saw the Smithsonian as a potential catalyst for supporting the scientific investigations that high-altitude rocketry would enable and a source of funding that should be pursued in tandem with military support. Events did not proceed exactly as Goddard expected, however. While Webster's appeal to the Navy Consulting Board generated only tepid interest, within three months of receiving Goddard's letter, the Smithsonian provided him with his first research grant.

A significant part of the enthusiastic Smithsonian response can be attributed to the engagement and advocacy of Charles Abbot—then the director of the Smithsonian Astrophysical Observatory and later the secretary of the Smithsonian—who would become one of Goddard's most ardent and long-standing supporters. Although technically a part of the federal government, the Smithsonian and its resources and charter were the result of Smithson's private bequest in 1835, and its operating structure was more similar to that of a private foundation than a government department.[24] Although it enjoyed less flexibility and more modest resources than some of the major private foundations, there was significant scope for individuals within the institution to promote research projects that interested them. Abbot was not only interested but evidently excited by Goddard's vision. He considered the concept quite plausible and had, in fact, independently discussed a similar idea with the eminent astronomer George Ellery Hale five years earlier.[25] Goddard's proposal thus appealed to Abbot's personal intrinsic interests. In correspondence with Charles Walcott, the secretary of the Smithsonian, Abbot strongly advocated support for Goddard's work. He provided the most extensive review of Goddard's manuscript, on the merits of which the Smithsonian awarded Goddard a grant of $5,000 from the Hodgkins Fund in January 1916.[26] It was thus through the private philanthropic legacies of two individuals, James Smithson and Thomas Hodgkins, that Goddard was first granted support. One year later, however, before this first grant had even been fully expended, Goddard—with the assistance of Abbot, the Smithsonian, and the declaration of war against Germany—would secure his second, much larger grant, this time from the military.

For Goddard, the entry of the United States into the First World War was an obvious opportunity, and he wasted no time in fostering military demand for his long-range-rocket technology. With no evident interest emerging from the Naval Consulting Board, Goddard continued to gently remind the Smithsonian of the military utility of his technology. Five days after the congressional declaration of war on April 6, 1917, however, he made a direct appeal:

> I feel that if the apparatus for reaching high altitudes, upon which I am working, has any military applications, these should be developed as soon as possible. Such applications may be of considerable value for these reasons:
>
> 1. Possibility of long ranges, exceeding that of artillery
> 2. Ease of transportation and of firing
> 3. No necessity for a fixed position of firing, with the consequent impossibility of the enemy's locating the firing position.[27]

Goddard then stressed the need for secrecy in the letter and boldly suggested that he be granted an officer's commission from the War Department as a ballistics expert "that would make my position sufficiently official to facilitate getting supplies and materials, as well as any information on ordnance that would be necessary during the construction of the apparatus."[28] As for the meteorological work supported by the institution, he felt it should of course be continued, "for such work would have an immediate military application in the attainment of very great ranges."[29] With the onset of war, Goddard had made the scientific applications of his invention subservient to its military applications, even in his discourse with scientists.

Abbot was as enthusiastic as Goddard about the military applications of the technology. He wrote Walcott, who was also the chairman of the influential Military Committee of the National Research Council, expressing his support in strong terms in April 1917:

> I believe he would succeed in perfecting means to propel large bombs or shells to distances of say 100 miles from the point of firing. The lateral aim I should think would be good, but the under- or over-shooting of the target quite probable. However if it were desired to destroy the Krupp works at Essen a large number of trials with different probable ranges might probably accomplish it from the French lines by means of Goddard's invention, provided the Allies were as willing to disregard the rights of noncombatants in Germany as are the Germans to murder noncombatants everywhere.[30]

He concluded his letter with a recommendation that a War Department grant of up to $50,000 to Goddard was warranted. Walcott, however, was initially less eager to promote the technology and replied to Goddard that "until demonstration has been made that the apparatus can be actually used, there will be very little upon which to base a request for the cooperation of the War Department."[31] Goddard initially showed deference to Walcott on the matter, but he was anxious for a more rapid pace of development. By the end of August, he independently sent a report to the chief of the Army Ordnance Department with the details of his Smithsonian grant and estimates as to the probable weights, payload capacities, ranges, and military applications of his rockets.[32] Although this report failed to convince the Ordnance Department to take immediate action, it was Goddard's independent pursuit of military support in the summer and fall of 1917 that ultimately led to military interest and funding.

Goddard's first supporter within the military was Halsey Dunwoody, a dynamic individual who would use his position within the recently expanded Aviation Section of the Army Signal Corps to advocate for Goddard's rocketry work. Although Lehman and Clary have downplayed Dunwoody's role, Goddard's diary entries make clear that Dunwoody was, in fact, a critical connection leading to Signal Corps support. Dunwoody would later rise to prominence during the war as chief of the Technical Section of the Air Service overseas in Paris and a member of the Interallied Aviation Commission. After the war, he was executive vice president of the Universal Aviation Corporation, where he established one of the first transcontinental train-air-train routes. He would also become vice president of American Airlines during the Second World War.[33] In late August 1917, however, he had just received his first commission as a lieutenant colonel in the Signal Corps when Goddard paid a visit to him at the U.S. Military Academy at West Point, where he was acting professor of natural and experimental philosophy.[34]

The ambitious Dunwoody immediately grasped the potential of Goddard's rocket designs. Although he likely did not yet have the formal authority to do so, he offered Goddard $5,000 in support.[35] Goddard seems to have been much encouraged by Dunwoody's enthusiastic response, writing, "I had about decided it was hopeless and thought I would give it to the best-looking man."[36] Dunwoody moved to Washington for his new role with the Signal Corps and resumed correspondence with Goddard in October, prompting Goddard's first of many trips to Washington on November 11 in search of funding.[37] While he met with Dunwoody twice during the visit, he did not meet with any of his contacts at the Smithsonian, and it is not clear if he even alerted them to his presence in the city.[38] The following month, Dunwoody visited Goddard at Clark to see his work, an impressive show of support when one considers Dunwoody's mounting responsibilities at the time.[39] By January 1918, Dunwoody seems to have prearranged for $10,000 from the Signal Corps. Goddard traveled to Washington on January 17 to meet with Walcott and Abbot for the first time to help write an official letter and accompanying technical report to Major General George Squier, the chief Signal Corps officer, requesting the funds.[40] With Dunwoody's groundwork and official recommendations of support from Walcott and Samuel Stratton, respectively the chairman and secretary of the National Research Council's Military Committee, approval was a mere formality.[41]

While Goddard's personal pursuit and cultivation of military support had enabled him to secure another grant for his rocket technology, there was a critical enabling factor on the demand side of the equation: the massive increase in military expenditure on aviation in the Signal Corps under General Squier. In July 1917, President Wilson had signed into law the largest individual appropriation for a single purpose in American history up to that point—$640 million for a new military aircraft program under the Signal Corps.[42] For comparison, the previous year's funding for military aeronautics had been $13 million.[43] The flood of funding created new opportunities for exchange between technologists and military strategists and an ideal context for Goddard's bold research proposal. The armaments-expenditure boom of the First World War had resulted in the first military funding for a long-range rocket technology that had been designed by its inventor with the intent of spaceflight.[44] Other very similar scenarios—proposals for long-range rocketry for bombardment, amid the rapid and massive increase in military expenditure during the armaments races before, during, and after World War II—would ultimately lead to the funding and successful realization of the spaceflight-capable liquid-fuel-rocket technology in Germany, the Soviet Union, and America.[45]

Along with new opportunities, however, the First World War funding deluge for aviation technology also created significant management problems. The new aviation production program was soon charged with widespread waste and corruption, with the result being that it was transferred out of the Signal Corps on May 21, 1918.[46] In spite of this, Chief Signals Officer Squier, a firm believer that science and engineering were crucial to the nation's economic and military competitiveness, was able to throw an additional strong line of support to Goddard. Squier authorized, less than a week before the program was moved, an increase of Goddard's grant to $20,000, allocating an extra $5,000 in August and working to subsequently encourage the support of the Army Ordnance Department.[47] Although the situation existed for only a relatively brief period of time, the First World War led to a major increase in American military expenditures and in the demand for advanced military technology—a demand that Goddard leveraged with opportunism and great determination in order to boost his spaceflight development program.

Wartime military budgets also led to Goddard's first encounters with industry interest in his rockets. In their first meeting, Dunwoody saw the potential for rockets to become big business given the recent increases in military expenditures, commenting, "can't you see that there's millions in that thing for you?"[48] Dunwoody suggested that Goddard partner with a manufacturer, since the government did not usually undertake these types of technology development projects in-house. Although Dunwoody offered to help put him in contact with potential partners, Goddard had his own ideas and initiative. He had already been developing a relationship with a major arms manufacturer—Winchester Bennett, president of the Winchester Repeating Arms Company. Goddard had been in touch with Bennett and the Winchester Company for a number of years, with the relationship developing to the point at which Bennett provided Goddard with steel for his experiments, free of charge.[49] In November 1917, Goddard visited the Winchester plant in New Haven, Connecticut, but was not able to engender a partnership—although Bennett did offer the Winchester firing range for trials.[50]

With his principal lead turning up cold, Goddard made enquiries with George I. Rockwood, president of Rockwood Sprinkler, a prominent Worcester automatic-fire-suppression-system contractor with $1.6 million in annual revenue.[51] Throughout December, Goddard and Rockwood discussed terms, toured each other's facilities, and developed a "tentative agreement" for the production of Goddard's rockets for sale to the military.[52] The agreement broke down in January 1918 due to disagreements over royalties and other rights. Rockwood, however, continued to see commercial potential in the rockets, and when relations soured between Goddard and his capable foreman Carleton Haigis, Rockwood immediately hired Haigis after Goddard dismissed him. Rockwood then pursued a separate rocket development contract with the assistance of an ambitious Ordnance Department officer, Colonel E. M. Shinkle. This resulted in significant consternation for Goddard and necessitated a formal letter from Walcott to Squier to deal with the situation.[53] Although

Lehman has claimed that Goddard's dealings with Rockwood resulted in an aversion to working with private industry, he in fact continued to pursue numerous industry collaborators throughout his career in connection with military contracts—including with Winchester again in 1919–1920.[54]

The immediate upshot of the Rockwood affair was that, in order to ensure greater secrecy and to improve productivity, Goddard's project was moved to California, where he could conduct his work at the Pasadena workshop of the Mount Wilson Observatory. The day after Walcott's letter to Squier, Walcott wrote to Robert Woodward, president of the Carnegie Institution of Washington, which funded the observatory, requesting that the facilities be made available, noting that all expenses would be borne by the government.[55] Goddard and his team relocated to Pasadena ten days later, on June 10, 1918. There, in the workshop where the observatory's famous one-hundred-inch telescope was still being completed, Goddard began pursuing two research projects under the aegis of the Signal Corps—a single-charge, tube-launched device, which would become a forerunner to the bazooka, and his multiple-charge, cartridge-based apparatus for attaining high altitudes, which he believed could scale to a system capable of reaching space.

Although Goddard had initially engaged the military because he believed it was the most likely source of funding for his multiple-charge, high-altitude rocket, the dynamic of this exchange led him to shift his focus. With the Signal Corps looking for rapid results, and with Goddard eager to please his new patrons, Goddard recognized that the simpler, single-charge device was his best chance for quick success. As a result, he was forced to allocate significant time and resources to it. The single-charge device correspondingly made rapid progress while, by the end of the summer, he was lamenting that "the multiple-charge devices have not received as much attention as desirable."[56] Nonetheless, he had assigned his most capable and creative assistant, Clarence Hickman, to the multiple-charge program, and he continued to tout its advances highly despite its more meager progress.[57] While Goddard had managed to leverage military resources for the development of his spaceflight technology, military demands had also leveraged Goddard's eagerness for support and co-opted him into more near-term and prosaic weapons development. The influence of military expectations on Goddard's resource-allocation decisions during the First World War is an example of the subtle change in objectives that can occur when spaceflight technologists, desirous of increased funding, enter into exchanges with patrons who do not share their intrinsic motivations.

From Goddard's perspective, the armistice of November 11, 1918, came at a particularly inopportune time. Goddard had worked intensively at Mount Wilson and within two months of arriving had almost expended the entirety of the initial $20,000 grant. On August 17, George Ellery Hale, the master fund-raiser and organizer of astronomical observatories, sent Abbot a note that gives particular insight into Goddard's approach to project management: "He is an enthusiast, and has been crowding the shop so much (against my advice) that his charges for overtime will be heavy. His funds will surely be exhausted by Sept. 1, and perhaps sooner, as he has a childlike way of forgetting about expenditures."[58] Although the $5,000 supplementary grant from Squier would keep Goddard working at Mount Wilson until the end of October, military enthusiasm for the development phase of the project was coming to a close. Goddard had, however, made strong progress and had impressed the Ordnance Department officers who had visited Pasadena in September to evaluate his devices. Arrangements were made for an official demonstration on November 6— of the single-charge and multiple-charge solid-fuel rockets, as well as of a double-expansion mortar gun he had developed—in front of a collection of Signal Corps, Army Ordnance, Navy Ordnance, and Aircraft Armament officers at the army's Aberdeen Proving Grounds in Maryland. While the multiple-charge device failed, the single-charge tests were successful, generating considerable interest from a number of the officers.[59] Captain Purinton of Aircraft Armaments, in particular, was enthusiastic about Goddard's recoilless rocket guns for aircraft. However, the end of the war, five days later, seemed to put into abeyance any potential demand for Goddard's rockets that might have been generated by the November 6 demonstration. Captain Purinton's comments to Goddard captured the sense of postwar institutional change: "Two months ago the thing would have gone through quickly. Saturday they stopped further development work, and planned not to take on new work until Dept. was re-organized."[60]

As the war ended, so too did Goddard's immediate prospect for using military demands to advance his plans for spaceflight technology.

Tellingly, however, Goddard continued his stubborn pursuit of military funding well after the end of the war. His intermittent work from 1920 to 1923 on rocket-propelled depth charges and antiship armor-piercing rockets for the navy at the Indian Head Proving Grounds—work for which he received some $2,000—is fairly well known to scholars of Goddard's career.[61] His encouragement of the potential gas-warfare applications of his rockets deserves particular investigation, however, as it has gone almost entirely unmentioned in previous histories.[62] He repeatedly suggested the application of his multiple-charge rocket to gas attack in letters to Abbot and in his reports to the Smithsonian.[63] Walcott and Abbot, in turn, encouraged Lieutenant Colonel Amos Fries, chief of the U.S. Chemical Warfare Service, to contact Goddard, which he did in May 1920.[64] Goddard responded immediately, suggesting a single-charge device for near-term gas-attack application and the possibility of a multiple-charge device for reaching ranges in excess of 3,000 yards (2.75 kilometers).[65] The two continued to correspond through June, with Goddard further suggesting the possibility of a long-range multiple-charge gas-attack device capable of 13,000 yards (11.8 kilometers).[66] In a letter of June 10, there was strong interest from Fries in the multiple-charge device, and he suggested that an officer could be sent to Worcester to discuss the matter with Goddard in detail. He also invited Goddard to visit the Chemical Warfare Service's Edgewood Arsenal at the Aberdeen Proving Grounds.[67] Goddard's diary entry of August 7 indicates there may have been action on the matter: "Had letter from CWS offered $25,000."[68] It is difficult to know if funding was ever produced, as the Goddard Collection at Clark University contains no financial documents or information related to Goddard's career other than what is contained in his correspondence and diaries. However, given that Goddard visited Edgewood Arsenal in November, continued to develop and test his multiple-charge rocket device with no other funding source through to January 1921, and wrote up a report for the Chemical Warfare Service in April 1923, it seems possible that some funding may have been provided.[69] It is also noteworthy that, in a letter to Goddard in November 1922, Major Adelno Gibson re-quested a meeting with the inventor, writing that "the Chemical Warfare Service has been working for some time on a rocket for use in carrying chemical warfare agents."[70] Goddard's letters and diary entries, and Esther Goddard's editorial comments, suggest that there may have been a Chemical Warfare connection with his work for the navy at Indian Head—work that he considered top secret.[71] Regardless of whether or not he received funding from the Chemical Warfare Service, it is clear that God-dard actively pursued rocket development for long-range gas warfare well after the horrific effects of gas attacks in the First World War were widely known and not long before international agreement was reached to pro-hibit their use in the Geneva Protocol, signed in June 1925.[72]

Goddard's pursuit of gas-warfare applications for his rockets, as well as his enthusiasm for weapons development during the First World War in general, provide strong evidence that a reevaluation of Goddard's moti-vations and objectives for his military research is in order. The list of weap-ons projects that Goddard pursued during and shortly after the First World War is extensive: rockets for long-range bombardment, portable single-charge rocket launchers, double-expansion mortar guns, recoilless rockets for aircraft, rocket-propelled depth charges, rocket-propelled anti-ship armor-piercing projectiles, and long-range delivery systems for gas warfare.[73] Although Goddard pushed most strongly for developments re-lated to the multiple-charge apparatus that he believed could lead to sys-tems capable of reaching orbit, he also allowed himself to be involved with projects with less-clear relevance to spaceflight. Given his pursuit of these after the war, wartime patriotism can only be part of the answer. He con-tinued to pursue defense-related funding directly and through his patrons throughout the interwar period, although he was not significantly suc-cessful in the effort until the armaments buildup prior to the Second World War, when he again worked for the military, this time on work closely related to the liquid-fuel rocket system that he believed would reach orbit. All this suggests that his enthusiasm for military research stemmed to a significant extent from his belief that military support was the most probable source of resources for the near-term development of spaceflight. That Goddard was willing to develop such a wide variety of weapons in pursuit of this support, including a weapon that could inflict gas warfare on victims over ten kilometers away, is sobering evidence of the Faustian bargain that Goddard was willing to make in pursuit of his singular and obsessive dedication to spaceflight.

On the whole, a reevaluation of the role of the First World War in Goddard's career and in the development of spaceflight in general also seems in order. Although it would be the Second World War that would enable the development of the first vehicle to reach above Earth's atmosphere, the First World War enabled the emergence of the first major rocketry research and development program motivated, at least on the supply side, by a spaceflight objective. As with the liquid-fuel rocket development led by Wernher von Braun at Peenemünde in Germany during the Second World War, the objective of spaceflight was not embraced by or even known to those that provided the funds to Goddard. It was nonetheless a principal motivation in his initiation of the project. Furthermore, while it was short-lived, Goddard's wartime research program was a high-profile one. It had the regular attention of and was supported by some of the most powerful people in American military research—Walcott, Stratton, Hale, and Squier. The project had access to top-class facilities and had the interest of multiple branches of the War Department. With an expenditure of $25,000 over roughly eight months—around seventeen times Goddard's annual salary as head of the Physics Department at Clark—it was an amply funded research-and-development project, given the stage of the technology at the time. The war also led to industry interest in the technology and taught Goddard his first lessons in private partnerships, government bureaucracy, and project management. Interestingly, in Europe, the First World War also seems to have stimulated plans for spaceflight technology development. Hermann Oberth, believing that the liquid-fuel-rocket designs he was working on could win the war by bombarding London, later wrote that he made his first pursuit of serious rocket development when he traveled to the German consulate in Kronstadt—now Brasov, Romania—to have his rocket designs sent to Berlin, only to receive a reply that deemed his idea impractical.[74]

In America, the imperative of war and the attendant weapons research budgets did allow a pioneer of spaceflight to conduct meaningful experimental work on his spaceflight technology. The potential military application of Goddard's rocket was even the hook he used when he pitched his story to Worcester's *Evening Gazette* after the war. On March 28, 1919, *The Evening Gazette* provided the first public exposure of his work, taking a distinctly militaristic approach in its first of many stories on Goddard:

> INVENTS ROCKET WITH ALTITUDE RANGE 70 MILES
> Terrible Engine of War Developed in Worcester by Dr. Robert Goddard, Professor of Physics at Clark, in Laboratory of Worcester Tech, Under Patronage of U.S. War Department.[75]

It would be through this story, which would be picked up by city newspapers across the country, and the first open publication of his research shortly thereafter, that Goddard removed the veil of secrecy under which he had been working. He did so in order to pursue other types of public support, this time leveraging direct interest in his rockets' spaceflight applications.

The postwar era was an important transition for Goddard. Although he was back in Worcester and no longer at the center of a major funded research program, his leave from teaching at Clark had been extended through the school year. While much of his time was dedicated to pursuing various military projects, he also had time to strike out in new directions. In particular, in January 1919, he resumed his considerations and calculations for liquid-fuel rockets, which he had effectively suspended in 1909.[76] The liquid-fuel rocket would eventually become his driving objective. In early 1919, however, the multiple-charge solid-fuel apparatus remained his focus and, with the profile he had gained from the *Evening Gazette* article, he decided it was time to go public with his plans. In a letter to Abbot on April 7, Goddard inquired as to the possibility of publishing under Smithsonian auspices an updated version of his original manuscript on the high-altitude solid-fuel rocket.[77] Abbot responded that there was still $3,000 remaining from the original $5,000 grant from the Hodgkins Fund and that the Smithsonian would agree to publish the work if Goddard would permit the use of some of those funds to cover the costs.[78] The manuscript would be published in the Smithsonian Miscellaneous Collections, volume 71, number 2, in December 1919 under the title "A Method of Reaching Extreme Altitudes" and would mark a major turning point in the development of spaceflight. In the meantime,

Goddard made his first and last attempt to start a business—again related to his pursuit of military dollars.

The company Industrial Research Laboratories of Worcester was initiated by Goddard along with two partners: Dr. Louis Thompson, a colleague at Clark who had worked on military applications during the war, and Nils Riffolt, Goddard's graduate student and long-serving machinist. The group had formed the company in May 1919, copurchasing a company car and printing a letterhead. Although Thompson seems to have been the principal motivator, Goddard evidently gave the concern significant attention. In a letter to Abbot in September, Goddard pointed to the company as the reason he had recently been spending only a small amount of time on the multiple-charge device: "The chief reason for this is the time and effort I have expended in helping start the 'Industrial Research Laboratories' of Worcester, in which I am one of the partners . . . a demonstration having just been staged at the laboratory for a large arms concern, with which we have already done considerable business."[79] Evidently this "large arms concern" was the Winchester Repeating Arms Company, as Goddard's diary records numerous instances of conversations with Thompson and Riffolt about a "Winchester proposition" and a visit of the three of them to New Haven in January 1920.[80] This visit, which was the last reference to the company in his diaries, also coincided with the increasing publicity that followed Goddard's Smithsonian publication. It is unclear what role either event played in the company fading from further reference in the diaries. The short-lived relationship again confirms Goddard's eagerness to be involved with armaments production, as well as his entrepreneurial streak and interest in establishing business partnerships as part of his plans for spaceflight. However, the failure of Industrial Research Laboratories, despite a strong stable of well-connected researchers and directors, is also an early indication of Goddard's limited ability to execute on his self-directed projects without the assistance and guidance of a patron.[81] This inability would come to the fore over the next two years as he struggled to convert the newspaper story of the year—his "moon-going rocket"—into a single instance of funding.

When the Smithsonian published Goddard's "A Method of Reaching Extreme Altitudes" in December 1919, the result was a genuine and sustained explosion of interest in spaceflight as the American public began to consider it a real near-term possibility. Although Goddard's manuscript was highly technical and conservative in its claims, his statements under the heading "Calculation of Minimum Mass Required to Raise One Pound to an 'Infinite' Altitude" generated significant publicity. The text that generated the excitement discussed principally the question of how to prove that a payload had reached an effectively "infinite" altitude: "The only reliable procedure would be to send the smallest mass of flash powder possible to the dark surface of the moon when in conjunction (i.e., the 'new' moon), in such a way that it would be ignited on impact. The light would then be visible in a powerful telescope."[82] This idea, buried at the end of the report, was highlighted in the accompanying press release from the Smithsonian. With a Smithsonian publication declaring the possibility of a rocket reaching the Moon, the result was predictably sensational. The *Boston Herald*'s headline was representative of many:

NEW ROCKET DEVISED BY PROF. GODDARD MAY HIT FACE OF THE MOON: Clark College Professor Has Perfected Invention for Exploring Space—Smithsonian Society Backs It[83]

On January 12, 1920, headlines of this type appeared in newspapers across the country, from New York to San Francisco. The following day, the excitement was even stronger, with articles now accompanied by spaceflight-themed political cartoons, opinion pieces, and letters.[84] Some were more sedate, such as the *Newark News,* which stated simply ATMOSPHERE EDGE CONDITIONS TO BE STUDIED BY ROCKET.[85] There was no question, however, which version of the story made better copy—or sold more papers. In an era with often multiple newspapers and multiple editions competing in a single city, exciting headlines and new updates were critical to journalistic success. Goddard's fantastic story of space travel provided the perfect hook, and the story continued to grow. On January 16, it was announced that Rear Admiral William S. Sims, commander of all U.S. Navy forces in Europe during the war, believed that it was possible that a rocket could reach the Moon as Goddard had described, adding further public credibility to the idea.[86] The public discussion of spaceflight soon made rapid strides in the pages of the nation's dailies. One such example is this meditation, originally from the *Cleveland Plain Dealer*: "And with this as a beginning who can tell how far we might not shoot in the centuries to come? Mars is the next step, and beyond Mars are untold myriads of worlds to conquer. One obvious difficulty in the way of shooting a star instead of the moon or Mars is that it would require a good many dozens of billions of years for the most active rockets to reach its destination, and the people of the earth might forget all about the experiment before its completion."[87] Spaceflight was hot off the presses and the nation was enthusiastic about a flight that was expected to occur in the very near future. Goddard's farsighted vision had stoked a popular demand for space exploration.

Goddard's first statement after the story broke, however, gives an early indication of his mixed feelings toward the public enthusiasm and his inability to convert this enthusiasm into financial support. He had been very cautious in his claims and had not even begun to enunciate his more ambitious plans and visions of interplanetary travel in the manuscript. He was evidently also somewhat uncertain as to how to react to the public attention he had generated. On January 17, *The Pittsburgh Dispatch* carried an overtly negative statement from Goddard under the heading NOT A MOON ROCKET: "While popular interest in the successful flight of a rocket from the earth to the moon may be intense, the scientific importance is so negligible that it could not justify the experiment."[88] In a curious juxtaposition, however, the following day he released a statement that initiated his request for a major fund of $50,000–$100,000. The scale of his funding request—at two to four times the size of his prodigious wartime military grant—shows his ambition. For comparison, the hottest new luxury at the time, a 1919 Model T Ford touring car, now with electric starter, cost only $525.[89] Amazingly, even when announcing his intention to raise a fund by popular subscription,[90] he continued to downplay expectations:

> In the first place too much attention has been concentrated on the proposed flash-powder experiment and too little on the exploration of the atmosphere. My reason for saying this is not because I believe the former is entirely unrealizable. In fact, if I were to speculate boldly instead of timidly I would say that based upon equally sound physical principles is the possibility of obtaining

photographs in space by an apparatus guided by photosensitive cells . . . precautions being taken to ensure a sufficiently safe and conspicuous landing on the return. To continue a speculation, however, on matters concerning which there is little experimental data would, of course, be unwise.[91]

Unwise, perhaps, in the eyes of a fastidiously cautious scientist, but for a person endeavoring to raise funds for a spectacular development, when the press was full of gusto and enthusiasm on the subject, to request a massive public subscription campaign in association with such negative and cautious statements demonstrated more naïveté than wisdom.

Newspapers writers were nonetheless impressively perceptive on how Goddard's funding request could potentially play out. *The Baltimore News* immediately saw the potential source from which Goddard's most significant funds would come: "The sum of $50,000 or $100,000 which Dr. Goddard suggests as necessary to get his rockets in working order is not large, as prices go nowadays. He will hardly get it by popular subscription, but millionaires have financed wilder schemes."[92] The focus on the potential interest of the wealthy in spaceflight would continue in the press for a number of years: "Wanted: A millionaire who is tired of this earth and would like to travel to the moon. There are many who would like to leave the earth, but they haven't the money to pay the fare."[93] On a more general level, *The Bridgeport Telegram,* in a witty piece discussing the initial issue of stock for a fictional company, "The Inter-Stellar Rapid Transit Co.," highlighted the potential legacy value of the endeavor: "Consider the chap who paid a comet discoverer $50,000 recently to act as godfather and namesake for the celestial newcomer."[94] The piece also explicitly recognized the nonmarket, and nonprofit, nature of the early development: "No, 'Cautious Investor' and 'Inexperienced,' no, this stock will not be for such as you. Stick to your oil and your punctureless tires and new-fangled fast mail pouch catchers and leave Inter-Stellar for those who can afford to invest in notoriety, even honor, without regard to profits."[95] It seemed to observers in the 1920s that intrinsic interests, signaling motives, and private financing would propel forward the exploration of space.

The popular view that private financing and the interests of the wealthy would be the most likely source of support for spaceflight already had cultural momentum prior to the public announcement of Goddard's ambitions. John Jacob Astor IV, scion and heir to the Astor family fortune, wrote an extensively researched work of spaceflight fiction, *A Journey in Other Worlds*, in 1894.[96] For the first time, an individual with the resources and inclination to invest in spaceflight expressed an eagerness to do so. For the first time, in Astor's book, the potential role of government support and technical assistance was recognized as a possible enabler for the development of a privately funded spacecraft as part of a public-private partnership: all previous American stories of spaceflight fiction had in fact depicted spaceflight as being achieved by individual will and resources alone. Most importantly though, *A Journey in Other Worlds* gave the idea of spaceflight a level of social acceptance that it had not previously enjoyed. The very fact that an Astor had written such a book lent the idea credibility. Though reviews outside of New York were more critical, those from within the burgeoning commercial metropolis were positively enthusiastic, with one declaring that "large public attention" had been centered on the book for weeks.[97]Astor's well-known book thus helped lay the groundwork for the almost wholly positive public reception that liquid-fuel rocketry pioneer Robert Goddard received when his work and ambitions became known to the world. Indeed, it is fascinating to consider what type of relationship the two men might have developed had John Jacob Astor IV not decided, after helping his wife and others to the lifeboats, to go down with the *Titanic* in 1912 at the age of forty-eight, and what the combined forces of the Astors, the Guggenheims, and Robert Goddard might have achieved for the development of American rocketry.

The initial public enthusiasm for the Goddard announcement was such that individuals were soon publicly declaring their intentions to travel on Goddard's rocket. Regular updates and repeats of the story continued through the month, and on February 5, Captain Claude Collins of the New York City Air Police announced that he was volunteering to fly to Mars.[98] The story gained widespread media coverage, and when Collins announced that all he requested in return was a $10,000 insurance policy for the benefit of his heirs in the case of accident, the predictable result was an immediate announcement by four others that they would be prepared to take the trip uninsured.[99] In total, in the next few years, over a hundred people would volunteer to travel into space on Goddard's rockets.[100] By the end of February, popular enthusiasm and expectations for interplanetary flight had become so great that *The Spokesman-Review* in Spokane, Washington, felt obliged to try to provide some basic logistical insight into the matter:

> THAT FLIGHT TO PLANET MARS—Would Take a Year to Reach the Goal and the Traveler Would Be Hungry.[101]

Although Goddard enjoyed the flood of publicity, the unending volunteering and lack of funding soon became a source of annoyance to him; he later commented that what he needed then was "less volunteering and more solid support."[102]

Goddard nonetheless actively tried to leverage this enthusiasm into a public subscription campaign. The fact that he chose this funding model, and the details of his ad hoc speaking tour, reveal his limited experience as a public fund-raiser. As recognized by *The Baltimore News,* the public subscription model had been in steep decline in the early twentieth century, eclipsed by the rise of private philanthropic foundations. Goddard had also long ago recognized the potential for single, wealthy individuals to undertake complicated, expensive engineering projects, as seen in his juvenile short story about a magnetic-levitation train project, in which he wrote of his wealthy protagonist's ambition, "now, where there are millions, there must surely be a way."[103] Despite this early recognition, as well as his significant connections through the Smithsonian, however, he eschewed a concerted campaign of personal approaches to potential patrons. Instead he placed his appeal to potential wealthy patrons within a general campaign for public subscription. In doing so, he was looking to employ the same model as had been used in Verne's *From the Earth to the Moon* and in some of the civic astronomical observatories of the previous century. Appropriately enough, the most publicized of Goddard's fund-raising speeches, to the Chicago chapter of the American Association of Engineers, was in the city that, over half a century earlier, had built the last of the great civic observatories funded by public subscription.[104] In a large hall, five hundred people listened to Goddard lecture on his plans and responded enthusiastically to them, resolving to assist him not by reaching for their wallets but by lobbying the millionaires of Chicago on his behalf.[105]

Although the resultant headlines—such as WINDY CITY CAPITALISTS MAY FINANCE WORK ON ROCKET TO REACH MOON and CHICAGO IS READY TO PUSH ROCKET TO MOON—as well as the secondary literature suggest that this was a significant event with serious promise, the poster for the evening tells a slightly different story.[106] For example, the largest text on the poster proudly proclaims that the event was LADIES' NIGHT at the Chicago chapter of the American Association of Engineers, with Goddard listed as part of the night's program of "Speakers—Movies—Entertainment—Dancing" and the bottom line imploring "Show Your Wife, Sister or Sweetheart What Chicago Chapter Is Doing For Its Members. EVERYBODY COME!"[107] This was not an auspicious occasion for a spaceflight fund-raising pitch, but Goddard was relatively inexperienced and public speeches were a rarity for him. That he would find himself engaging in his most fervent fund-raising talk at such an event is telling of his inexperience and the difficulty he would have in navigating the world of public fund-raising.

A further factor complicating Goddard's efforts was his tendency to overpromise before he had acquired the resources to deliver. Through April, Goddard pushed hard to find support, visiting polar explorer Admiral Byrd and Alexander Graham Bell, who, in addition to developing his own inventions, had been a significant early supporter of aeronautics.[108] He also visited the National Geographic Society with Abbot and, while there, worked with them to make an ill-advised public announcement that a launch would occur on the Fourth of July—despite not having secured any funding.[109] The results were headlines reading GODDARD ROCKET TO SHOOT IN JULY followed by embarrassing postponements, first to August and then indefinitely.[110] It does not seem to have occurred to Goddard that, in announcing that a launch was already forthcoming as a fait accompli, he was limiting the interest of people who might have been interested in funding the event.

Although the press coverage of Goddard's "moon-going rocket" was extensive—four folders are required in the Clark Archive for Goddard's press clippings from 1920—there is almost no mention of the military implications of Goddard's rocket. This is because Goddard, whose repeated interviews seeded the information for the articles, did not mention them publicly during this time. Although it was known that Goddard's rocket had been developed under the patronage of the War Department during the First World War, Goddard spoke solely of its peaceable uses and the press largely followed his lead.[111] Even when he became desperate for funding, he continued to maintain that the only value of the rocket was to science.[112] Why would he not have discussed the valuable military implications of his rocket when trying to raise funds publicly? One explanation is that, since Goddard's objective was always spaceflight, with military funds only a means to that end, perhaps he simply hoped that the world's enthusiasm for space travel might be adequate motivation to assure the necessary funding support. It had worked for Jules Verne's Baltimore Gun Club. The American Rocket Society would later labor under a similar misconception—with George Pendray noting that he and the society believed "that a few public meetings and some newspaper declarations were all that would be necessary to bring forth adequate public support for the space-flight program."[113]

Another more subtle reason for Goddard's reticence to speak of military applications in his immediate postwar funding campaign is hinted at in one of the few negative commentaries that appeared in the press, entitled GODDARD'S ROCKET IN WAR. In it, the author found it "rather surprising" that there had been so little "discussion of uses in war for Prof. Goddard's multiple-rocket" given that it would "make nothing at all of the earth's distances" and "with it continents could make war on each other from their own shores"; the article ended with a hearty "on with the war-averting league of nations!"[114] The Great War had left the world reeling from its atrocities, and there was little enthusiasm to contemplate further conflict so shortly after its end. If Goddard had publicly tried to highlight his rocket's military applications in 1920, as he frequently did in private, he might well have been labeled a warmonger and his bid for public funding would have been almost certain to fail. He would have to wait until American enthusiasm for armaments again increased before he could publicly discuss the military implications of his invention. Until that happened, he would continue to state that the sole applications of his rocket were scientific, in an effort to raise funds on that merit alone.

By September of 1920, Goddard felt he had exhausted his options and was candid with the press about his inability to solve the financial problem of spaceflight. To *The Washington Times* he confided, " 'To tell you the truth,' said Prof. Goddard, 'I have been terribly cramped for money. The Smithsonian Institution backed me for $5,000 but that is practically exhausted . . . and I really do not know where to look for the greater sum which will be necessary to make a definite trip to the moon.' "[115] Goddard identified what he thought was the problem in a feature piece entitled ONLY "ANGEL" NEEDED FOR TRIP TO MOON: "It would cost a fortune to make a rocket to hit the moon. But wouldn't it be worth a fortune? The great pity is that I cannot commercialize my idea. If I could rant of a 100 percent return in forty-five days, I'd have been financed long ago. But, unfortunately, the proposed rocket is worth little save to the world of science."[116] With no "angels" visible, the newspapers were soon trying to defensively reassure an interested public that "that earth-to-the-moon scheme has by no means been abandoned."[117] Eventually, however, even the Worcester *Evening Gazette* had turned on Goddard, flatly declaring in a headline that GODDARD'S ROCKET IS NOT PRACTICAL.[118] As it became clear that no Moon trip was imminent, general interest in Goddard and his Moon rocket subsided until the subject was almost out of the press altogether. Months later, however, a small group of "angels" finally appeared—in the form of Clark University's president Wallace Atwood and its Board of Trustees.

In June of 1921, Clark University provided Goddard with a critical grant of $2,500. This support—along with another $1,000 grant the following year and a salary increase to $2,500 plus $1,000 for assistants, apparatus, and supplies—was of critical importance for Goddard, as it enabled his final tests on the multiple-charge rocket and allowed him to pursue the development of the liquid-fuel-rocket engine on which he would work for the remainder of his life.[119] Goddard had discussed the matter of support with Atwood earlier in April and prepared a report for the board with a supporting letter from Walcott. Goddard's application laid specific emphasis on the public recognition that would come to the university from his endeavors—effectively appealing to the desire to signal Clark's commitment to quality research: "In view of the fact that this work promises to establish a new era in investigations at very high altitudes, and also that it was begun experimentally and has been continued for the greater part of the development at Clark University, it is believed that this request to the Trustees is not out of place, considering the recognition which will come to the University on the completion of a satisfactory demonstration."[120] The result was both short-term and long-term recognition, the first coming in the form of the morning and evening headlines: CLARK TO FINANCE GODDARD'S ROCKET: EXPERIMENTS WHICH DREW WORLDWIDE ATTENTION TO GO ON WITH THE UNIVERSITY'S AID.[121] Although he continued with solid-fuel-rocket experiments for a short time at Clark and Indian Head, the numerous difficulties with the system progressively led him to withdraw from the solid-fuel line of development in favor of the liquid-fuel rocket. By the time of the first grant report to Clark in April of 1922, he had decided to devote his entire attention to this new direction.[122] The age of the liquid-fuel rocket was close at hand.

Allowing that the liquid-fuel rocket may have already been close at hand in 1922, it was the publication of Hermann Oberth's *Die Rakete zu den Planetenräumen* in 1923 that would make it seem to the world as if it was within reach. There is good reason to consider *Die Rakete zu den Planetenräumen* as, in the words of Arthur C. Clarke, one of the few books "that have changed the history of mankind."[123] In contrast to Goddard's understated rhetoric and minutely technical work, Oberth began his work with four bold propositions:

1. With the present state of scientific knowledge and the science of technology, it is possible to build machines that can rise higher than the limit of the atmosphere.
2. If perfected further, these machines can attain speeds by virtue of which—if left to themselves in ether space—they do not have to fall back to the earth's surface again and are even able to leave the sphere of attraction of the earth.
3. Such machines can be built so that human beings (apparently without danger to health) can go up with them.
4. Under today's economic conditions, it will pay to build such machines.[124]

The impact of Oberth's work was enormous. Along with being heavily technical and mathematical, it was also significantly more literary and expansive than Goddard's initial publication. Oberth also actively encouraged the boldest of public spaceflight speculations, with the result that a significant spaceflight movement soon developed within the Germanic-speaking countries.[125]

Goddard's reaction was instant animosity, as Oberth had boldly made public many of the long-held ideas that Goddard had been purposefully keeping private, notably the importance of liquid-fuel rocketry. Goddard remained obsessed with "that German" for decades, convinced that Oberth had stolen his work. The publication of *Die Rakete zu den Planetenräumen,* however, was in fact a significant boon to Goddard's development program. The year 1923 had started out as a difficult one for Goddard, as Clark had ceased the grants for his rocket research and he was forced to continue with his liquid-fuel work on his own salary and lab money. The publication of Oberth's book in May, however, generated a new wave of media attention for spaceflight, this time including concerns over the possible advantage that recent-enemy Germany might have in the technology. Goddard looked to use this to his advantage in a report—which for the first time included some of his more ambitious plans for interplanetary travel—that he sent to Clark and the Smithsonian, in which he played on the prospect of a space race with Germany: "I am not surprised that Germany has awakened to the importance and the development possibilities of the work, and I would not be surprised if it were only a matter of time before the research would become something in the nature of a race."[126] Abbot, however, was not especially moved at the time, given his now almost decade-long familiarity with Goddard's hype and promises. He replied to Goddard that he was "consumed with impatience, and hope that you will be able to actually send a rocket up into the air some time soon. Interplanetary space would look much nearer to me after I had seen one of your rockets go up five or six miles in our own atmosphere."[127] Nonetheless, the renewed media attention had rejuvenated some interest, and Goddard secured two grants in 1924: a small grant of $190 from the American Association for the Advancement of Science (AAAS), and $500 from a new private fund—the Cottrell Fund—provided to the Smithsonian by prominent inventor and philanthropist Frederick Cottrell.[128]

Dr. Frederick Cottrell, a major figure in the development of American philanthropic support for technology research, has received little attention in the Goddard historiography, even though the use of the Cottrell Fund to support Goddard's research is an important example of the culture of private philanthropy that enabled Goddard's research. Cottrell's biographical memoir, written by Vannevar Bush, the great science administrator and organizer of the Manhattan Project, focuses especially on the innovations of the simply titled organization that provided the funds to the Smithsonian—the Research Corporation.[129] The Research Corporation was unique at the time, at once a business and a nonprofit entity. It was created by Cottrell and his associates donating to it their patents in the field of electrical precipitation, with the proceeds from the applied research to go to the further advancement of science and technology. It operated as a business based on the patents, but it paid no dividends to personal stockholders. Instead, its income above expenses, apart from reserves and operating capital, was expended to assist scientific and educational institutions in carrying on research.[130] In 1924 the Research Corporation gave the Smithsonian $5,000 to support technological innovation, and, over the course of the next three years, with the personal approval of the president of the Research Corporation, A. A. Hamerschlag, all of it would be allocated to Goddard's liquid-fuel-propulsion research.[131] The Research Corporation, the most unsung of Goddard's supporters, also provided him with an additional $2,500 in 1929, after the success of his flight of the first liquid-fuel rocket. Although it took a more hands-off approach than Goddard's other funding sources, its contribution came at a critical time, and it was with its funds that the development leading to the flight of Goddard's first liquid-fuel rocket was conducted.

As with many milestones in space exploration, that of March 16, 1926, when Goddard's liquid-fuel rocket first left its test stand, was a simple, small step. Shielded from the distractions of publicity by his supporters at the Smithsonian, Goddard had been productive under the new grants. He reported with pleasure that he had his liquid-fuel engine working steadily in the shop in April 1924, and by March 1926, with snow still on the ground, he was ready to make his flight.[132] At his Aunt Effie's farm in rural Massachusetts, Goddard, his wife, and two assistants prepared the apparatus and recorded the event with photographs and film. Although it has been ensconced as a historic moment, his diary entry for the day was as plain as any: "Tried rocket at 2:30. It rose 41 ft, and went 184 ft, in 2.5 sec, after the lower half of nozzle had burned off. Brought materials to lab. Read Mechanics, Physics of Air, and wrote up experiment in evening."[133] The following day, however, he allowed himself a little more description: "It looked almost magical as it rose, without any appreciably greater noise or flame, as if it said, 'I've been here long enough; I think I'll be going somewhere else, if you don't mind.'"[134] This first small flight, and the comparable ones that shortly followed, were warmly received when reported to Abbot at the Smithsonian. Abbot remained impatient, however, for a genuine high-altitude flight. When he asked what was needed to make that happen, however, Goddard gave a serious underestimate, suggesting that at least $2,500 would be needed.[135] Whether this was because Goddard knew that precisely $2,500 was left in the Cottrell Fund, which was allocated to Goddard after his note, or because this was his honest estimate is unclear. Regardless, with further private support, out of the limelight, Goddard's development of the liquid-fuel rocket continued.

The continued Smithsonian impatience for a high-altitude flight proved to be another example of how Goddard's development program was influenced by the requirement to make exchanges in order to receive funds. In a move that may have been spurred by an ongoing $10-million fundraising effort by the Smithsonian, Abbot upped the pressure in responding to Goddard in June 1926: "We do not feel like embarking $2500 more in this business unless we can see our way pretty clearly to a certainty that it will lead to a high spectacular flight. . . . Hitherto, for the past two or three years, we have been supplying additional sums of $500 at a time with the expectation that each in succession would bring the matter to a climax. That is just what we do not wish to do any longer, but to go about the matter with a well-grounded assurance that it is going to bring it to a fruition, or else drop it right here."[136] This prompted Goddard to begin working on a new rocket that was twenty times larger than his earlier efforts. It was too much too soon, and after a year of work, in the summer of 1927, Goddard was forced to admit his mistake and scale down the rocket to a more manageable size, expending a substantial amount of time and effort in the process.[137] In the meantime, the pressure on Goddard continued, with a direct request for a spectacular show of his rocket in conjunction with the Smithsonian fund-raising campaign: "There is no use in premature tests, but if it should happen that a successful flight was made before February 11, it would be a grand thing for the Smithsonian Institution. On that date we have a conference at the Institution, to be attended by the President of the United States, the Cabinet, the Regents, and about forty guests of the highest eminence, to consider plans to sustain and further endow work of the Smithsonian Institution." The Smithsonian had understood the rocket's signaling power and looked to Goddard as Khrushchev would later look to Korolev: after the success of the first Sputnik, Khrushchev had asked Korolev whether another satellite could be launched in time for the celebration of the fortieth anniversary of the Great October Socialist Revolution. Even in the early days, before a major flight of a liquid-fuel rocket had been achieved, signaling demands were beginning to influence the pace and direction of rocketry development.

By 1928, the time Goddard had wasted on the large rocket meant that, although he was now making progress on the scaled-down version, his funding situation was again dire. He resumed his search for other sources of support, turning once again to the potential military applications of his invention. He enlisted Abbot to make an appeal to Congress for $10,000, citing German interests in the military development of rocketry.[138] Although the appeal was unsuccessful, Abbot still provided some support to Goddard—making two extraordinary grants that year of $250 and $1,500 from the Smithsonian's own operating funds. Goddard's salary had also been increased to $4,000. This kept the work going but was not enough for the development program that Goddard envisioned.[139] He noted in his diary at the end of 1927 the costs of Commander Byrd's expedition to the North Pole ($240,000) and Antarctica ($500,000) and the sources of private support that had underwritten the expeditions—including Edsel Ford, Vincent Astor, John D. Rockefeller, and Rodman Wanamaker.[140] He also entertained the prospect of engaging industry and, in the process, formed an interesting personal connection to the Apollo program. In September, one of Goddard's former graduate students at Clark had contacted him about the undefined potential interest of his company in Goddard's work. This student was the aviation manager at the Standard Oil Development Company, Lieutenant Edwin E. Aldrin, whose son, Edwin E. Aldrin Jr., known more popularly as "Buzz," would be born two years later and would become the second man to walk on the surface of the Moon.[141] Although the records don't show a through line from their correspondence to Goddard's subsequent work, Goddard nonetheless expressed in his correspondence to Aldrin his appreciation of the role of private industry in the development of spaceflight systems: "I know perfectly well that the hardheaded businessmen, who, after all, are really the ones who put research developments on a going basis, are convinced only by final accomplishments, and are not influenced by theories alone, however sound they may be."[142] Although he had yet to achieve such a "final accomplishment," Goddard was continuing to engage private industry in preparation for the day when he expected to be able to put spaceflight on a "going basis."

In December, Abbot was also corresponding with Arthur Noyes at Caltech, which, under the leadership of George Hale, had just secured the funding for the Palomar Observatory and its two-hundred-inch reflector. Abbot reported to Goddard on December 27 that Noyes had "become much interested in the matter from what I told him last summer, and it may be that he and Dr. Hale may see their way to assist in the completion of the development in case you still see good hope in that direction."[143] Given Hale's familiarity with Goddard's research from their time together at Mount Wilson, and Hale's boyhood interest in spaceflight, it is tempting to think that Hale was considering one last great space exploration project. In late January 1929, Abbot sent a letter to Noyes officially requesting support from the fund for the two-hundred-inch telescope: "If, therefore, you should still be as favourably interested as you were last summer and should see your way toward raising for this purpose $20,000 or $25,000 and inviting Dr. Goddard to undertake these interesting studies in the neighborhood of Pasadena, I think it would be very well worth while."[144] After receiving a response, Abbot sent Goddard a letter that reveals what is possibly one of the most significant missed connections in space history. Abbot reports that Noyes discussed the possibility with Caltech and Mount Wilson staff and determined that "Although they were warmly interested in the project as he was himself, they felt that the terms of gift of the money for promoting the 200-inch telescope could not fairly cover the promotion of the Goddard rocket. Dr. Hale had been in Europe and unavailable, and would not be returning until some date in May."[145] It is impossible to know whether Hale's presence in Pasadena at the time of the letter would have made any difference. The collection of Hale's correspondence at the Caltech Archives contains no further information on the matter, so it does not seem that Hale had given the subject much consideration. However, given that Hale was without peer as the most accomplished space exploration fund-raiser of the era, and that Goddard was at the time without peer in the technical development of liquid-fuel rocketry, it is an interesting thought experiment to consider how different the history of spaceflight might have been if Hale had been present and had decided to take up the development of Goddard's rocket.

While these letters were being sent back and forth across the continent, Goddard continued to work toward his next model of rocket. By the summer of 1929, his efforts had achieved spectacular, headline-grabbing success. The sudden public drama of the flight of July 17, 1929, was captured by the Boston Globe:

> "MOON ROCKET" MAN'S TEST ALARMS WHOLE COUNTRYSIDE:
> BLAST AS METAL PROJECTILE IS FIRED THROUGH AUBURN TOWER
> ECHOES FOR MILES AROUND, STARTS HUNTS FOR FALLEN PLANE,
> AND FINALLY REVEALS GODDARD EXPERIMENT STATION[146]

Goddard had conducted a number of test flights since his early ones in 1926, but this was his first widespread national attention since the early days of 1920–1921. The rocket had risen 20 feet above the 60-foot launch tower and ended up 171 feet from the center of the tower: it had flown at an average velocity of about 55 feet per second.[147] The immediate upshot of this accomplishment was a torrent of newspaper men attempting to see his testing station. This resulted in Goddard and Abbot requesting and obtaining permission from the War Department to relocate the experiments to Camp Devens, near the Cambridge train station and twenty-five miles from Goddard's workshop.[148] Goddard was now firmly into productive development work and became somewhat annoyed by the renewed media attention. Nevertheless, it would have a direct positive impact on his funding. While he continued his development program under an additional $5,000 grant from the Smithsonian and the Research Corporation, the dramatic newspaper stories of the test flight had already put in motion a series of events that would lead to the most significant source of funding of his career—funding that for the first time would allow him to dedicate himself full-time to the development of liquid-fuel rocketry.[149]

The most significant funding for the early development of liquid-fuel rocketry in America did not come from the government or the military, as Goddard had expected, but from the private purses of the Guggenheim family. Goddard received funding directly from Daniel Guggenheim— who had built the family mining empire—and, after his death, from the Daniel and Florence Guggenheim Foundation with the assistance and advocacy of Daniel's son, Harry Guggenheim. As Pendray tells the story, Harry Guggenheim, an avid aviation enthusiast and supporter, and his close friend Colonel Charles Lindbergh, then one of the most famous men in America after he had been the first to complete a nonstop transatlantic flight, were discussing the problem of high-altitude aviation on a July day at Falaise, Harry's home in Port Washington, New York. While the two men were discussing the problem, Harry's wife, Caroline, who was reading the paper, noticed a story about Goddard's recent rocket flight.[150] Harry suggested to Lindbergh that this may be the answer to high-altitude flight and suggested that he get in contact with Goddard. Although the story does not account for the four-month delay between when the events occurred and when Lindbergh telephoned Goddard on November 22, 1929, the support from Lindbergh was immediate and strong once contact was made. Lindbergh visited Goddard's laboratory and home on November 23, and the two began their long conversation about the future of spaceflight. When Lindbergh asked whether it would be possible to send a rocket to the Moon, Goddard responded in the affirmative but added that it would cost at least a million dollars.[151] Goddard then turned the conversation to what he thought was a more reasonable program—$25,000 a year for four years to develop a high-altitude rocket. Lindbergh would later write of the encounter in his autobiography: "the thought of sending a rocket to the moon set my mind spinning."[152] Lindbergh was hooked and quickly worked to find Goddard his funds.

Interestingly, however, the first connection Lindbergh made for Goddard was not to a philanthropic source but to private industry—the Du Pont Company, which had once held a monopoly over the American explosives market. Lindbergh asked Goddard to meet him on November 29, at the office of Henry du Pont, head of the DuPont Company. During the ensuing meeting, three DuPont laboratory employees, who Goddard said had been looking into the question of liquid oxygen rockets "in a very general way," grilled him about the technical details of his rockets and their near-term applications.[153] This evidently made both Goddard and Lindbergh uncomfortable, and, after personally flying Goddard back to Worcester, Lindbergh apparently confided in him, saying, "The Du Pont people did not have the right attitude towards the work. . . . He said that any extensive support should come from some person interested in the scientific side of things rather than in their immediate applications, and said he himself was greatly interested in the work, was convinced of its importance, and thought it ought to receive very substantial support even on the basis of the result already attained. He said he knew Dr. Merriam of the Carnegie Institution very well, and intended to see him in a few days."[154] Thus Goddard was taken under the wing of one of America's most prominent figures and given privileged access into the world of the rich and powerful. A new field of patrons and opportunities soon opened up before him.

Within four days, on December 2, Abbot wrote to Goddard of a visit he had received from John Merriam, president of the Carnegie Institution, who shared a desire to take Goddard's work to a new level: "Dr. Merriam indicated that he thought that $100,000 or more should be made available and, as I understand him, Col. Lindbergh is likely to see the way to raise this money for the purpose." On December 10, a conference was convened at Merriam's apartment in Washington, with Merriam, Goddard, Lindbergh, and Abbot, as well as Drs. Walter Adams and Horace Babcock from the Carnegie Institution–funded Mount Wilson Observatory, Dr. Charles Marvin, chief of the U.S. Weather Bureau, and Dr. John Fleming of the Department of Terrestrial Magnetism of the Carnegie Institution.[155] Goddard's report from the meeting shows the various scientific interests that had converged on the technology: Lindbergh's interest in the rocket's relevance to aviation; Adams's interest in using the rocket to perform spectrographic analysis of the Sun above the atmosphere; Fleming's ideas on the possibility of using the technology to measure the magnetic field lines and ion content above the atmosphere and to investigate ionospheric radio propagation; and Babcock's thoughts on the possibility of using the vehicle to take solar coronagraphs without waiting for an eclipse.[156] When Marvin estimated that $100,000 would be required to cover ten miles and that $500,000 might be necessary for a hundred-mile-capable system, Lindbergh stated that he was willing to devote all the time he could to furthering the work and that he was sure he could secure the necessary funds.[157] After two hours the conference split up, and Goddard emerged with a new nucleus of support centered on Lindbergh, who had become enamored with both Goddard and the vision of spaceflight.

The Carnegie Institution's support was the first to become material, although Lindbergh and the Guggenheim family were not far behind. On December 19, Merriam sent a letter to Goddard offering him $5,000 in funding—a direct match of the Smithsonian's last grant and an evident attempt to put the Carnegie Institution on equal footing with Goddard's longest-standing institutional supporter. On June 12, 1930, however, a letter was delivered to Atwood at Clark that would change Goddard's fortunes more substantially.

Although Lindbergh had not mentioned the Guggenheims at the conference, he knew that he could count on their interest and support if he asked for it. The immense mining fortune of the Guggenheim family had established the Daniel Guggenheim Fund for the Promotion of Aeronautics, and ten Guggenheim Schools of Aeronautics sprang up across the country. These had already played a major role in the development of passenger aircraft, scheduled airliners, and general aeronautics. The philanthropy had also given a high-visibility boost to the family's reputation, which had suffered in previous decades due to the family's connections to bloody strike-breaking activity, political bribery, and unethical labor policies.[158] With Harry in Cuba as the American ambassador, Lindbergh visited Daniel Guggenheim personally and argued that rocket flight was a logical extension of aeronautics and that Goddard was the man to extend it. Convinced of the endeavor's worthiness on Lindbergh's recommendation, Daniel Guggenheim wrote to Atwood offering Goddard $50,000 in funding over the next two years, along with a statement to the effect that if progress on the work was deemed satisfactory, another $50,000 would be made available.[159] This was the start of what would become the most significant single source of support for Goddard's space-flight development program. It was an arrangement that would make Goddard one of the most richly financed independent American scientists in the early twentieth century outside of a U.S. government or corporate laboratory. Although family signaling motivations were part of the general backdrop to the Guggenheims' philanthropy toward scientific research, their support of Goddard was essentially a reflection of the intrinsic interests of a small group of wealthy and influential individuals who were interested in high-altitude flight and who shared in Goddard's own personal motivations.

The initial grant was roughly equivalent to all the grants that Goddard had received over the previous twelve years combined. Atwood immediately accepted the offer and established at Clark the Daniel Guggenheim Fund for the Measurement and Investigation of High Altitudes. In late July, after the school year ended, Goddard would move to Roswell, New Mexico, with his wife and team, which a year later would consist of five technicians and machinists. There, on a secluded farm, Goddard established his workshop and test range. Over the next eleven years, he would develop his liquid-fuel rocket through 103 static tests and 48 flight tests, incorporating essentially all the technical elements that were being integrated, more or less contemporaneously, into the V-2 rocket (for example, curtain cooling, gyroscopic stabilization, turbopumps).[160] Over those eleven years, the Guggenheim family would contribute $188,500 to Goddard's research, which would allow him to dedicate himself almost exclusively to the development of his rocket. Even with the death of Daniel Guggenheim in October 1930, his widow, Florence Guggenheim, continued the initial grant on the recommendation of Harry and Lindbergh.[161]

The commencement of the "Guggenheim Rocket Research Project," as Goddard called it, has largely been interpreted by the secondary literature as marking the end of Goddard's search for funding until the start of the Second World War. In fact, the Guggenheim funding was never enough to satisfy Goddard's ambitions, and he continued to search for other sources of funding, specifically from the military. Shortly after arriving in Roswell, he reinitiated contact with the Army Ordnance Department and received two delegations of officers from the department in

1931 and 1932.[162] He did not tell his sponsors back east about either of these visits. In 1932, the Great Depression temporarily required the Guggenheim family to suspend their support for Goddard, forcing him to return to Worcester and to turn to his old patron, Abbot, who managed to find another $250 dollars in the Hodgkins Fund. Such small funding made Goddard acutely aware of the effect that the economic crisis had on private funding opportunities: "To my mind, the future of such a new scientific development as the engineering side of rocketry will depend largely on how soon, and to what extent, recovery is made from the depression, which is at present unpredictable."[163] Military funding, if it could be made consistent, again seemed preferable to him.

Goddard contacted Lindbergh in May 1933 to notify him of his shifting interests: "It appears that the rocket will have applications as an antiaircraft weapon, owing to high speed and controllability. . . . Incidentally a work of this kind could be well camouflaged in New Mexico, with easy access to Government officials, through the New Mexico Military Institute. . . . If you can spare the time, I would very much like to discuss the matter with you, at your convenience, together with Mr. Guggenheim and Colonel Breckinridge, if they should care to be present."[164] Lindbergh, still recovering from the shocking kidnapping and murder of his son a year earlier, did not respond. Goddard took his case to Abbot, who arranged for meetings with the navy in late June, where he met Dr. George Lewis, of the National Advisory Committee on Aeronautics (NACA), as well as Admiral King and Commander Pownall. The navy expressed interest in his rocket "air torpedo" concept and told him to expect a letter from the secretary of the navy.[165] Goddard soon received a letter from Secretary Henry L. Roosevelt, to which he responded with a long report outlining two "possible applications to national defense"—the use of his rocket as a power plant and as an air torpedo. After more letters and review, Acting Secretary Admiral W. H. Standley informed Goddard on August 29 that "because of the great expense that would be entailed in the development of the rocket principle for ordnance and aircraft propulsion, which under present stringency of funds appears hardly warranted, the Department regrets it is not in a position to further such development."[166] Goddard, while pursuing the navy, had also asked Harry Guggenheim to renew his grant of $25,000 for work at Roswell, or at least $2,500 for continuing work at Clark. Harry committed the $2,500 to support his work at Clark in 1933 and was soon able to again promise $18,000 from the Daniel and Florence Guggenheim Foundation for the research to be continued again at Roswell in 1934.[167] For the next seven years, support of $18,000–$20,000 was provided annually until Goddard decided to pursue military contracts with the start of the Second World War.

Although the Guggenheim funding proved to be a steady stream, it was never guaranteed as such. The Guggenheim grant had to be renewed on an annual basis, with the result that Goddard was forced to deal with perpetual funding uncertainties—uncertainties that interfered with both research at Roswell and pedagogy at Clark. Goddard continually felt that "a continuance beyond this year would depend quite as much, or more, on financial conditions as on the results attained."[168] His sponsors did, however, consider results important, specifically high-altitude ones. In a letter to Goddard in 1936, Lindbergh stressed the uniquely fortunate situation of having an advisory committee and sponsor that appreciated the long-term potential of Goddard's rockets, but felt the need to add: "I feel that the morale of everyone concerned would be greatly increased if you find it possible to obtain a record-breaking flight."[169] Although this seems an understandable request, it is important to put in context. The statement comes, in some respects, at the height of Goddard's progress, having recently achieved a flight of 2.3 kilometers in altitude. For comparison, the flight of von Braun's A-1 in 1934 was only 1.7 kilometers. Whereas von Braun subsequently received a massive infusion of capital after such a successful flight—the commitment from the Wehrmacht Ordnance and the Luftwaffe to build Peenemünde—the response by Goddard's supporters was further encouragement to reach higher altitudes with the same amount of funding.[170] The "record" being referenced was the altitude record for sounding balloons, which at the time was some 20 miles. That Lindbergh and others expected flights an order of magnitude higher than Goddard's current capabilities with no additional increase in funds is a testament to the unrealistic expectations that they held for the work. Though Goddard may have fueled these expectations by deliberately understating the resources required to reach such a height, they were yet another reason for him to continue his search for military funding throughout his years at Roswell.

When Goddard received a letter expressing interest in a rocket-propelled drone for antiaircraft target practice from Major General A. H. Sunderland, chief of the Coast Artillery Corps, a little over a year after returning to Roswell, he jumped at the opportunity. In October 1936, Lieutenant John W. Sessums Jr. of the U.S. Army Air Corps flew out to visit and evaluate the Goddard workshop for Sunderland and was inundated with a flood of enthusiasm and information from Goddard. "Goddard was so absorbed in telling me about the status of his work, and the future possibilities of it, that he drove in first gear the entire fifteen miles from the launch tower back to the shop."[171] There was no immediate funding result, as Sessums's army air corps superiors were not as enthusiastic as Goddard or Sunderland. Nonetheless, although Lehman suggests that the relationship with Sunderland ended there until the Second World War, the two in fact continued to scheme about ways of getting the military to fund the concept, both in correspondence and in meetings with Sunderland in Washington, right through to the start of the war.[172]

Understanding Goddard's continued pursuit of military support helps to explain what Clary referred to as "the most persistent mistake of his career"—his emphasis on gyroscopic stabilization for his rocket.[173] Clary attributes this mistake to an assessment that Goddard "was not trying to build a guided missile, which would have required steering to reach a destination. Rather, his were sounding rockets, intended to rise straight up and then return to Earth by parachute. Gyroscopes could be useful in steering, but were cumbersome and unnecessary for stabilization; properly placing the rocket's center of gravity forward would suffice, but that also was a ballistic principle that escaped him."[174] In fact, the evidence suggests that Goddard was indeed trying to develop a guided missile. This evidence can be found in the letters and reports that he wrote to both Roosevelt and Sunderland, in which, trying to sell his program, he prominently placed statements emphasizing that he and his team had "carried out a sufficient number of tests to demonstrate practical flight control by gyro, both during and after propulsion, and also parachute release and descent."[175] Clary faults Goddard for not taking the simplest and most effective approach to his task of the moment—a high-altitude sounding rocket. But Goddard was not focused only on that objective. He was also thinking and planning for the next step in the development of spaceflight technology, which he rightly knew would be guided missiles.

By September 1938, with the threat of war in Europe looming large, Harry Guggenheim, who had also been funding rocketry work under Theodore von Kármán at the Guggenheim Aeronautical Laboratory at the California Institute of Technology (GALCIT), had come to the same conclusion about the need for guided missiles. He had been helped to reach this conclusion by Goddard, who, in his letters to Guggenheim and Lindbergh, had increasingly begun to emphasize the military applications of his rockets, as war seemed increasingly imminent. Lindbergh came to similar conclusions during his tour of German aviation capabilities for the American government. He wrote to Goddard in July 1937, "It seems to me that the first practical use for liquid-propelled rockets may be for military purposes. . . . There is constantly increasing interests in rockets among military men in Europe, although it is confined to comparatively few officers at present. Don't you think it might be advisable to establish closer contact with our Ordnance Department?"[176] Goddard hardly needed the encouragement, responding to Lindbergh that "I feel, as you do, that the rocket is inherently an offensive rather than a defensive weapon. It is likely, however, that applications to antiaircraft problems will be made first, because of the great need for developments of this sort. This does not, of course, alter the desirability of developing offensive applications, such as long-range shelling and high-speed planes." Lindbergh and Goddard continued an extensive correspondence on the military applications of rocketry and the best ways of interesting the military in their development.[177] Goddard also addressed the subject squarely with Guggenheim in his annual report in February 1938: "In view of the present state of world affairs it seems desirable to continue the work along the most advantageous lines, especially since the problem has been brought to a point where definite applications appear to be within reach, and I would be glad to help in any way I can to have military applications realized as early as possible"[178] In September 1938, Guggenheim called together his entire informal rocketry advisory board, including Goddard, von Kármán, Professor Clark Millikan of Caltech, and NACA representatives. He informed them that they "were all engaged in working for the country's defense" and that they should henceforth pursue more active work on military applications.[179] Goddard had now been given completely free rein in his engagement with the military. Within a few years, he would secure from the navy the single largest annual research budget of his career.

Goddard's final years working for the navy on jet-assisted takeoff (JATO) rockets, and the personal difficulties he encountered in doing so, are well known. What is not so well known is how aggressively he pursed this new role and how he made a conscious decision to leave the auspices of the Guggenheim Foundation in favor of the military. Lehman explains the move as an initiative of Guggenheim's, stating that Guggenheim informed Goddard that he "planned to offer the professor's services and the New Mexico establishment to the military services," and that "Goddard had accepted the proposal."[180] Clary suggests that Goddard's turn toward the military stemmed from the 1938 meeting at Falaise, after which he felt that further foundation support would not be forthcoming.[181] The evidence does not support this position. The Guggenheim Foundation would continue to support Goddard fully for another two years and would provide Goddard extraordinary support during the transition to military work in 1941, including a $3,000 grant and a $10,000 loan, repaid in 1943.[182] Goddard had little reason to worry about a stoppage in funding, as Guggenheim had also recently vowed on national radio to support Goddard's work "indefinitely."[183] There is no indication that Guggenheim wanted to off-load his rocket researcher to the military. On the contrary, while supporting Goddard's military engagement, Guggenheim was also concerned that it might distract him from his main purpose of high-altitude flight. Guggenheim wrote to Goddard in July 1941 stating, "We do not want to jeopardize the future of the work we have already carried on, unless there are some very good reasons from a National Defense point of view to do so."[184] The evidence argues more strongly for the view that Goddard's impetus for his work with the army and the navy in the Second World War was internally driven and stemmed from his long-held belief that the military was the most probable source for the large-scale funding needed to achieve his dream of spaceflight-capable rocketry.

In 1938, after the meeting at Falaise, Goddard increased the intensity of his lobbying for military funding. He wrote to Guggenheim enthusiastically noting that "problems regarding national defense are beginning to arise, both along the lines of aviation and ordnance, which require for their solution effective liquid-fuel rocket propulsion."[185] He continued his correspondence with General Sunderland and begun working with officers in charge of *Army Ordnance* magazine to insert a promotional article on liquid-fuel rocketry.[186] On his own initiative and with the help of Lindbergh, Guggenheim, and Abbot, his military contacts and connections soon proliferated. After a discussion with Guggenheim in October 1938, Major James "Jimmy" Doolittle, famed military flyer and aviation pioneer, flew out to Roswell to evaluate the wartime business potential of Goddard's work for his employer, the Shell Oil Company.[187] A second inquiry from industry came in early 1939 from ordnance inventor Lester P. Barlow, consulting engineer and special assistant to Glenn Martin at the Glenn L. Martin Company. The company expected to receive $100,000–$200,000 from Army Ordnance for glider-bomber development. They hoped to pursue rocket work in tandem with this effort and wanted to hire Goddard as a consultant.[188] Goddard eagerly pursued this relationship until it was clear that there would be no significant funds attached to the consulting work. He encouraged his former employees, many now in the armed services, such as Homer Boushey, Clarence Hickman, and Nils Riffolt, to lobby their respective departments for support. By 1940, Goddard was corresponding and meeting with a wide variety of high-profile figures within the military establishment, including General Henry "Hap" Arnold and Vannevar Bush.[189] As he had done during the First World War, Goddard again situated himself within the military research community at the highest level in an effort to leverage his expertise and access into major funding for his program.

Although Goddard had been self-catalyzing in approaching the military, his success was attributable in no small measure to a source of support that has been underappreciated in the Goddard historiography: the National Advisory Committee on Aeronautics (NACA) and its director of Aeronautical Research, Dr. George Lewis.[190] In a number of respects, the NACA was the initial locus of wartime support for Goddard, with Vannevar Bush as its chairman, Lindbergh and Guggenheim as influential committee members, and having previously included Charles Martin and Charles Abbot. Goddard's diary records that, on a visit to Washington on May 24, 1938, he met with the NACA's first employee and only-ever executive secretary, John F. Victory.[191] In June, Goddard reported this to Lindbergh, who responded in August that he had received a letter from George Lewis at the Langley Aeronautical Laboratory asking Lindbergh for Goddard's "recommendations with reference to any rocket research on which the National Advisory Committee for Aeronautics might focus."[192] Goddard responded with suggestions for liquid-propellant rockets on gliders, application of rockets in turbines, and rockets for accelerating and decelerating planes.[193]

The exchange was the start of a long and productive correspondence between the two researchers. Lewis wrote directly to Goddard in January 1938, offering the use of the Langley high-speed wind tunnels for aerodynamic tests of Goddard's rocket. This led to the testing of a model of one of Goddard's eighteen-inch rocket flight casings later that year.[194] Lewis's most significant assistance, however, came in June 1941, when he confided in Goddard that evidence from recently captured German planes showed evidence of JATO attachments. He suggested that if Goddard could submit a workable design for liquid-oxygen rocket-assisted takeoff, he could see a way for such a project to be initiated under the NACA or the navy's Bureau of Aeronautics.[195] The secondary literature has credited Boushey and Robert Truax with Goddard's navy contract, but although they played their roles, it was ultimately the word of the eminent Lewis that had the greatest impact. Goddard sent Lewis a report on June 23, and by July 14, Rear Admiral J. H. Towers, chief of the Bureau of Aeronautics, wrote to Goddard to notify him of the bureau's intention to offer him a contract.[196] Lewis and the NACA, the organization that would later be transformed into NASA, had led Goddard to his long-sought-after large military contract.

The JATO development contract with the navy would initiate the last major phase of Goddard's research and life. Its importance has been underappreciated, in part because its full financial value has never been put into context.

Although Goddard's lobbying of General Arnold had also paid off with a six-month, $13,000 contract from the army air force for JATO development, it was the navy project, starting at $40,000 for 1942, that would occupy him until his death in 1945.[197] Goddard worked on the initial six-month contracts at his workshop in Roswell, but by July he had moved to the U.S. Navy Engineering Experiment Station at Annapolis, Maryland, where he more than doubled the size of his team and workshop. As Goddard had foreseen in 1915, military development programs, once mobilized, could provide for large budgets. In July 1943, the navy exercised an option on his contract, extending it another year and increasing its value from $87,267 to $191,867—making it by far the single largest contract or grant of his career.[198] Military funding had, as he had long expected, enabled him to scale up his operations. As had been the case previously, however, the nature of his exchange with the military also shaped his objectives and development program—away from high-altitude rocketry and toward more-terrestrial, in fact horizontal, uses.

Although a JATO development contract may not seem like a move closer to spaceflight, it is important to recognize that Rhodes's evaluation of the situation was essentially correct: there was no U.S. military need for long-range rocketry in the Second World War.[199] When Goddard presented to Army Ordnance in 1940, the issue was not whether or not he made an effective presentation. The military leaders had been correct in their assessment of the near-term potential of rocketry and rightly concluded that there was no near-term need for rocketry in the U.S. armed services—with the sole possible exception of small rockets to boost airplanes on takeoff.[200] Moreover, unlike Germany, America did not have major military and civilian targets within a couple of hundred miles of its territory. Nor did it have a desperate dictator that was willing to gamble vast sums of money on an expensive and impressive, but ultimately strategically unimportant, psychological weapon. The rational evaluation by the U.S. armed services of the limited use of liquid-fuel rocketry for them within the context of the Second World War meant that there was only minimal potential for an exchange with the military to begin with. By this measure, Goddard showed significant skill in maximizing that potential—receiving major funds from the army air force and the navy.

There is a strain of argument—started by von Kármán—that suggests that it was the liquid-fuel rocketry work being done at GALCIT, rather than by Goddard, that was the important American rocketry program in the Second World War. It is worth noting, however, that there were only four major liquid-fuel rocket efforts during the war: the German program led by von Braun, the Russian Gas-Dynamic Laboratory and *sharashka* work led by Korolev and Tikhonravov, the JATO work at GALCIT, and the U.S. Navy program led by Goddard. Von Kármán at GALCIT did have greater success with the JATO problem, but largely because he and Frank Malina had set about trying to solve the specific problem presented to them by the military—emphasizing solid-fuel rockets and using red-fuming nitric acid and gasoline for their liquid-fuel rockets—whereas Goddard was trying to shoehorn the liquid-fuel rocketry approach that he believed could get him to spaceflight—liquid-oxygen and gasoline—into the military contract parameters. That Goddard was able to get his ambitious liquid-fuel JATO rocket program funded at all is in fact nothing short of remarkable.

Goddard had to sacrifice much, however, to achieve this feat: leaving the comfort of Guggenheim auspices, expending $4,500 of personal funds to keep the work going during the time of transition, and shifting away from his beloved high-altitude rockets to the simpler, if more powerful, rockets required for JATO applications.[201] It had also greatly increased his stress and replaced the warmth and solitary comfort of New Mexico with the political and climatic volatility of Maryland. The tragic result was that during the three years at Annapolis, as the responsibilities, bureaucratic paperwork, and stress mounted, Goddard's health declined rapidly, ultimately resulting in his death at age sixty-two, on August 10, 1945. Through great effort and sacrifice, Goddard had finally achieved the large military support that he had originally believed was required to develop his high-altitude rockets, only to be unable to see the full fruits of that final partnership.

Even as he was dying, however, Goddard was preparing to scale up his development program with his first and last real industry partner—the Curtiss-Wright Corporation, heir to the Wright brothers' corporate holdings and Goddard's employer at the time of his death. His objective was the advancement of his long-range goals for his liquid-fuel rockets, planning to turn over to Curtiss-Wright the task of manufacturing the JATOs and thereby developing an infrastructure that would enable him to carry on the development of his rocket after the war.[202] In 1942, he licensed a number of his rocket patents to Curtiss-Wright in the hope that they would protect his intellectual property from infringement and that the royalties, half of which were to be given to the Guggenheim Foundation, might be used to further advance his work.[203] As the war ended, a number of industrial concerns approached Goddard for assistance with their nascent liquid-fuel rocketry programs that, with the capture and reverse-engineering of the German V-2s, they hoped would become big business. In early 1945, Goddard met with Lovell Lawrence of Reaction Motors Incorporated, and with William Coolidge and Irving Langmuir of General Electric, with the result that both companies tried to attract him as an engineering consultant for expected upcoming rocketry work. Goddard remained committed to Curtiss-Wright, however, and just before his untimely death had made preparations to move himself and his team to the company headquarters in Caldwell, New Jersey, so as to focus on developing the capabilities required to compete for more military rocketry contracts in the postwar period. It is tempting to think that, had he lived into the Cold War era, Goddard would have taken up a prominent role working with the military and captains of industry, as he had so often done before, as America entered the Space Age.

An empirical examination of the funding sources tapped during Robert Goddard's career-long spaceflight development program provides a new perspective on both Goddard and the broader economic trends of early spaceflight technology. A chronological chart of the funding is provided in table 3.1. The dollar-value conversions are in terms of the production worker compensation (PWC) ratio, a metric useful for comparing the cost of projects, such as space technology development, where the chief input is skilled labor; and the relative share of GDP ratio, a metric that indicates what an equivalent share of total U.S. GDP would be worth today.

Contrary to the traditional conception of Goddard having been financed almost entirely by private philanthropists, military funding represented a large portion of the funds that he received to develop his rockets. Depending on the comparative metric used, military funding in fact provided some 75 percent of the level obtained from Goddard's more lauded private benefactors. Nonetheless, the majority of Goddard's funding

Table 3.1. Robert Goddard's Funding, 1917–1945

Year	Source	Nominal value in U.S. dollars ($)	Constant-price value in U.S. dollars ($) adjusted by PWC index, base year 2015	GDP-ratio equivalent value in U.S. dollars ($) adjusted by ratio to GDP, base year 2015
1917	Smithsonian (Hodgkins Fund)	5,000	536,000	1,500,000
1918	U.S. Army Signal Corp	25,000	2,130,000	5,890,000
1921	Clark University	2,500	158,000	607,000
1922	Clark University	1,000	68,800	243,000
1924	Smithsonian (Cottrell Fund)	5,000	301,000	1,030,000
1924	AAAS	190	11,400	39,000
1928	Smithsonian (Operations)	1,750	103,000	321,000
1929	Smithsonian (Research Corporation)	2,500	148,000	431,000
1929	Smithsonian (Operations)	2,500	148,000	431,000
1930	Carnegie Institution of Washington	5,000	290,000	978,000
1931	Daniel Guggenheim	50,000	2,980,000	11,700,000
1932	Smithsonian (Hodgkins Fund)	250	17,100	75,800
1933	Guggenheim Foundation	2,500	173,000	788,000
1934	Guggenheim Foundation	18,000	1,040,000	4,860,000
1935	Guggenheim Foundation	18,000	1,010,000	4,370,000
1936	Guggenheim Foundation	20,000	1,110,000	4,250,000
1937	Guggenheim Foundation	20,000	970,000	3,880,000
1938	Guggenheim Foundation	20,000	955,000	4,130,000
1939	Guggenheim Foundation	20,000	955,000	3,860,000
1940	Guggenheim Foundation	20,000	912,000	3,510,000
1941	Guggenheim Foundation	3,000	124,000	418,000
1942	U.S. Army Air Force	13,000	462,000	1,410,000
1942	U.S. Navy Bureau of Aeronautics	87,267	3,100,000	9,480,000
1943	U.S. Navy Bureau of Aeronautics	104,600	3,260,000	9,290,000
Totals	Private Sources	217,190	12,010,300	47,421,800
	Military	229,867	8,952,000	26,070,000
	Total	447,057	20,962,300	73,491,800

Sources: Goddard, R., *The Papers of Robert Goddard*, vol. 1 (New York, 1970);
see text of chapter 3 for specific source references. PWC-ratio equivalent value
and GDP-ratio equivalent value calculations done using measuringworth.com.

was still provided by private sources—a fact that the broader narrative of spaceflight history, with its focus on the heavily government-funded space race, has yet to fully incorporate.

The extent of the private philanthropic support that Goddard drew upon is quite impressive. No less than five separate bequests provided him with funding: those of James Smithson, Thomas Hodgkins, Frederick Cottrell, Andrew Carnegie, and Daniel Guggenheim. None of these mostly deceased philanthropists had been personally interested in space-flight technology. Even Daniel Guggenheim, who provided the funds while still alive, was motivated to do so principally out of the interest of his son Harry and the persuasion of Charles Lindbergh. This underscores the critical importance of well-connected patrons in Goddard's career. Goddard had a variety of such patrons that provided him with institutional support and connections to funding: Charles Walcott, Charles Abbot, Wallace Atwood, Charles Lindbergh, Harry Guggenheim, and George Lewis. Not only was Goddard's progress driven forward by his own intrinsic interests, he was also able to connect with patrons who either shared his interests directly or who saw alignment of their own intrinsic interests with what Goddard was seeking to achieve.

As we have already seen in the history of the American Observatory Movement, the private funding attracted by Goddard is not an anomaly in the history of space exploration. On the contrary, Goddard's program can be seen as part of a continuum of private funding for American space exploration going back for more than a century. Both as a share of total economic resources and in terms of equivalent production worker compensation, Goddard's career funding falls far short of the level of funding lavished on many of the nineteenth- or early-twentieth-century observatories. Nevertheless, at over $70 million in 2015 GDP-ratio terms, Goddard received over the course of his career a level of funding not dissimilar to what might be required for a nontrivial NASA technology development program of the twenty-first century. Or to put it into a more historical context, the next privately funded liquid-fuel rocketry program to receive similar funding in the United States would not occur until the first private launch-vehicle efforts of the 1980s.

The fact that Goddard's project funding, while substantial, was far below what could be obtained by American observatories can be attributed in part to the limited signaling ability of Goddard's early research and development program. With his rockets at a prototype phase and with so many fits and starts to his experimental programs, Goddard was unable to provide the type of signal that could command the private funds he required: given the stage of his research, it was often enough of a signal of far-sighted beneficence for his institutional funders at Clark University and the Smithsonian to fund him at all. When signaling did become a motivating concern, as in Abbot's impatience for a spectacular and newsworthy flight, patrons could be easily disappointed.

Nevertheless, the notion that a wealthy benefactor would come along to support the ambitious professor's rocket to the Moon was a common assumption among the American public—who had come to expect such an arrangement from the spaceflight fiction of the previous century—and indeed one did. As had been expected by the early American intellectuals of spaceflight, the most significant financial support for Goddard came from private-sector individuals who shared with Goddard a deeply felt intrinsic desire to explore the limits of flight. The support that Goddard received from Harry Guggenheim and Charles Lindbergh was substantial—but not substantial enough. Though the long-awaited start to American spaceflight technology development had arrived—and the project funded, as expected, by a philanthropically minded, wealthy American—the funding that emerged was nowhere near what was required for the realization of high-altitude liquid-fuel rocketry. As Goddard knew from an early age, the armaments budgets of the Western powers were a far easier and more generous source of funding for technological research if one was prepared to develop weapons—and to achieve his dream of spaceflight, Goddard was.

The extent and enthusiasm of Goddard's pursuit of military funding was significant. As a result of his efforts and exertions to interest the military in rocketry development, the first military funding that was used for space-flight development research came not in the Second World War, as is commonly believed, but during the First World War—with Goddard's pioneering multicharge solid rocket development program conducted in secret at the workshop of the Mount Wilson Observatory. After the Great War, Goddard continued to enthusiastically and consistently pursue military funding, including for the development of gas warfare technology under the Chemical Warfare Service. In all, he worked with and lobbied for support from a good portion of the technical branches of the U.S. military, including the Army Signal Corps, Army Ordnance, Navy Ordnance, Navy Aeronautics, and the National Advisory Committee on Aeronautics. Although the bulk of Goddard's military funding came only late in his career, in 1942 and 1943, it was a very substantial part of his overall program support. Nor did Goddard's pursuit of weapons-related funding support stop with the government. He pursued with equal vigor, if less success, numerous private-sector technology companies with interests in military sales: the Winchester Repeating Arms Company, Rockwood Sprinkler, DuPont, Glenn L. Martin, Curtiss-Wright, and General Electric. In short, Goddard was a man whose own intrinsic desire for spaceflight, and a future among the planets, was so deeply felt that he was willing to try every avenue he could to make it happen. In the early twentieth century, it simply seemed to him that the most promising avenues available were the ones that led to war.

Empirical matters and economic lessons of history aside, we have also seen that the willful pursuit of military funding extends back to the very start of the history of American spaceflight. There was no halcyon age when the leading spaceflight pioneers did not look to increasing military budgets to search out mutually beneficial relationships, particularly in wartime. The road to the stars was paved, if not necessarily with blood itself, then at least with a sincere willingness to improve bloodshedding capacity in the service of the nation. Robert Goddard, in both his technical acumen and his lust for military funding, was thus the equal of his great German competitor, Wernher von Braun, for they were competitors, on opposites side of the Second World War and in the direct competition for that great goal of Goddard's life—the development of a high-altitude rocket. The strategies they employed to achieve that goal were shaped by a shared belief—that, as von Braun put it, "the Army's money was the only hope for big progress toward space travel."[204] As with Wernher von Braun, the life of Robert Goddard thus serves not only as an inspirational story of technological vision and persistence but as a cautionary one as well, one that should remind us of some of the dangers that can arise from an obsessive dedication to spaceflight.

The military interest in delivering payloads long distances, which Goddard had worked hard to cultivate, would become overriding after the development of the nuclear bomb dramatically increased the potential for devastation that could be encapsulated in a "payload." Henceforth, space technology would become inextricable from matters of national defense, and the military demand for it would create a nationwide industrial base for advanced aerospace technologies from which almost all subsequent spaceflight and space exploration projects would benefit. At the origin of American liquid-fuel rocketry technology, however, there was simply an individual who desired to make spaceflight a reality. He had committed to dedicate his labor to its development over the course of many decades, and he had worked to create exchange and patronage relationships with the people and institutions that he thought could help him realize his near-term goal of a high-altitude rocket flight—a goal that was one step in his long-term objective of the navigation of interplanetary space, which in turn was part of an ultralong-term vision of a human migration to other star systems.

With this perspective in mind—and after examining in detail Goddard's career as the first practical developer of spaceflight technology and its first significant fund-raiser—it becomes possible to understand the development of space exploration in simplified terms. Space exploration as an economic outcome can be thought of as the result of the interaction of people who have intrinsic motivations for space exploration, who either self-support their efforts, if they have the resources to do so, or who enter into exchanges with others—with political, military, commercial, philanthropic, scientific, or other space exploration interests—on some basis for resources in order to pursue their space exploration objectives. Robert Goddard's career and space technology development program conforms to this pattern, as does much of the history of astronomical observatories covered in the previous two chapters. We turn now to that most famous example of space exploration—the Cold War space race—where the producers of space exploration entered into an economic exchange of unprecedented size and created the technology for human spaceflight.

4

IN THE EYES OF THE WORLD: THE SIGNALING VALUE OF SPACE EXPLORATION

One can predict with confidence that failure to master space means being second-best in the crucial arena of our Cold War world. In the eyes of the world, first in space means first, period; second in space is second in everything.

—*Lyndon Johnson, April 1961*

According to the conventional history of spaceflight, the Space Age began in 1957 with the launch of a small metallic sphere with radio antennae into orbit around the Earth. What had begun as the private ambition of a small cadre of individuals became a geopolitical event of major significance. Over the next decade, the two superpowers of the planet, the governments of the United States of America and the Union of Soviet Socialist Republics, would compete with each other to be the first to exhibit increasingly impressive, complex, and costly abilities in space in what became popularly known as the space race. The race into space resulted in an explosion of robotic probes out into the solar system, the first voyages of humans to the surface of the Moon, and the first long-term space habitations. In the history of technology, there are few events that match the Space Age in terms of political and cultural significance. In economic terms, the era remains unequaled in the amount of resources that were allocated to those engaged in the exploration of space.

As an epochal event, the dawn of the Space Age has engendered an extensive historical literature. Far from being underinvestigated, the period suffers from the opposite problem: the competing and complementary narratives are so numerous that the event has become overdetermined. This is particularly true for the American space program, which has borne the full force of academic analysis. The growth and evolution of the American space program has been portrayed as a result of Cold War competition for prestige, as a matter of national pride, as a military necessity, as part of the rise of technocracy, as a "Second Reconstruction" program for the American South, as the realized goal of spaceflight revolutionaries, as incremental policy development by the NASA bureaucracy, as an evolutionary inevitability for expanding life, or as all of the above and more.[1] One might question the value of adding another voice to this cacophony. Yet it argues by its sheer dissonance for another attempt to sort through the differing viewpoints and attempt a consolidated analysis of the exchange mechanism by which spaceflight activities acquired their extensive funding in the Cold War context.

The participants in the formulation of post-Sputnik American space policy referred consistently to "prestige" in justifying their funding requests. It is therefore not surprising that the concept of prestige-based competition has attained a preeminent explanatory position in the subsequent historiography. This has been the case since the earliest attempt at comprehensive analysis in Vernon Van Dyke's 1964 work *Pride and Power*.[2] The concept has since retained its prominent position, from Walter McDougall's Pulitzer Prize–winning account, *The Heavens and the Earth: A Political History of the Space Age,* in 1985, through to John Logsdon's comprehensive new summary, *John F. Kennedy and the Race to the Moon,* published in 2010.[3] Although prestige, a political concept popular in the nineteenth and early twentieth centuries, can be a useful frame for understanding the political motivations behind the space race, there is a modern concept from economics and biology that provides more explanatory heft—the concept of signaling, which emphasizes the ability of spaceflight activities to credibly transmit information about those undertaking them.

This chapter argues for understanding the competition of the Cold War space race, and the history of spaceflight more broadly, within a signaling framework. Such an approach acknowledges the force of prestige in space history but enfolds that concept within a broader conceptual framework, one that widens the explanatory scope and is also able to explain important historical events that do not fit the "prestige competition" thesis, such as Kennedy's willingness to cooperate with the Soviets to go to the Moon, and the evolution of the Space Station Freedom program into the International Space Station. Understanding the space race in the context of signaling also provides a new perspective on the domestic and international reactions to Sputnik, on Eisenhower's unsuccessful attempts to limit what he saw as a "prestige race," on the growth of space programs outside of the United States and the Soviet Union, and on the decreasing American political interest in spaceflight in the later part of the twentieth century. A signaling analysis also helps to explain how the legacy of the Apollo program has shaped and misshaped American space policy and created popular but misleading perceptions of U.S. space capabilities and the organizational competencies of NASA.

In order to appreciate the political value of spaceflight as a signaling device, we must disaggregate the political demands for space technologies from other applications—such as for military, commercial, or scientific purposes. In parsing this problem, it is helpful to refer to the taxonomy presented in one of the founding documents of the U.S. space program. In the historical narratives that search for the social, political, and economic forces behind the creation of NASA and the U.S. space exploration agenda of the 1960s, pride of place is rightfully given to the 1958 "Introduction to Outer Space" report. It was prepared by Eisenhower's Presidential Science Advisory Committee (PSAC) under James Killian's leadership. In it, four motives are listed as to why space technology is a priority, and these deserve to be quoted at length:

> The first of these factors is the compelling urge of man to explore and to discover, the thrust of curiosity that leads men to try to go where no one has gone before. Most of the surface of the earth has now been explored and men now turn to the exploration of outer space as their next objective.
>
> Second, there is the defense objective for the development of space technology. We wish to be sure that space is not used to endanger our security. If space is to be used for military purposes, we must be prepared to use space to defend ourselves.

Third, there is the factor of national prestige. To be strong and bold in space technology will enhance the prestige of the United States among the peoples of the world and create added confidence in our scientific, technological, industrial, and military strength.

Fourth, space technology affords new opportunities for scientific observation and experiment which will add to our knowledge and understanding of the earth, the solar system, and the universe.[4]

The assertion of the Science Advisory Committee that the first of the factors is the deeply felt need to extend ourselves into the universe suggests an important phenomenon often overlooked in the political history of space exploration. As we have seen, this intrinsic interest to explore has been identified as one of the most important motivations behind the dedication of successive generations of Americans to the challenges of spaceflight. As we have also seen with the stories of John Quincy Adams, George Ellery Hale, Robert Goddard, and others, the role of the individual is central to this. It is not so much that humanity feels a compelling urge to explore and discover, but rather that *some individuals* are determined to go boldly where no one has gone before, that *some individuals* turn to the exploration of outer space because of a primordial urge to do so. What this means, in economic terms, is that there is a segment of the population that is motivated to supply their intellect, capital, and labor for space exploration projects with limited sensitivity to the monetary reward for doing so.

Of the other three forces, I would argue that only two of them can rightfully be considered motives for space exploration per se. The prestige, science, and military motives have all received significant attention in the secondary literature. However, if we are examining specifically the field of spaceflight and space exploration, as opposed to space activities more generally, only prestige and science, in addition to the satisfying of intrinsic preferences, can be considered as direct objectives. As we have seen with Goddard, there is a close and even symbiotic relationship between military motives and space exploration. The military has gained a number of important capabilities from the efforts of engineers dedicated to space exploration, while those in pursuit of the planets have often leaned on the military in order to acquire the resources for their projects. Fundamentally, however, military objectives have not yet been themselves a motive for space exploration; thus far, they represent a separate need and service with their own separable demands.

That said, we must acknowledge that the military expenditures on space-related activities have often been dominant in the context of the U.S. government. General Bernard Schriever, often considered the "father of the U.S. missile program," believed "the compelling motive for the development of space technology is the requirement for national defense."[5] In 1959, the House Select Committee on Astronautics and Space Exploration concluded that "outer space is fast becoming the heart and soul of advanced military science. It constitutes at once the threat and the defense of man's existence on earth."[6] Senator John Stennis, chairman of the Senate Committee on Armed Services 1969–1981, put it more bluntly: "Space technology will eventually become the dominant factor in determining our national and military strength. Whoever controls space controls the world."[7] More than fifty-five years after these statements and prognostications were made, strong cases can be made for them. Although detailed public data on military space budgets is scarce, it is evident that military space programs overtook expenditures on government civilian space programs in America in 1982 and, with the exception of the period from 1994 to 2001, have continued to remain higher. In 2012, total Department of Defense (DoD) space budgets, including the National Reconnaissance Office, the National Geospatial-Intelligence Agency, the Missile Defense Agency, the air force and other DoD forces, totaled $26.7 billion, at a time when NASA's budget was $17.7 billion.[8]

As Senator Stennis predicted, space technology has become a key asymmetric capability that the United States and other global powers seek to establish and maintain in order to enhance global reach and power. Satellite communications allow real-time communication with deployed forces; satellite remote sensing enables strategic and situational awareness as well as effective identification of tactical threats; Global Positioning System satellites allow precise knowledge of force deployment and logistics, as well as precision guidance for munitions to within one meter. These technologies have augmented American war-fighting capabilities: from the first uses of the CORONA and GRAB reconnaissance satellites in 1960, through the Vietnam War, where defense meteorological satellites were first used to coordinate bombing campaigns; from the effective use of communication satellites in the First Gulf War, to the introduction of GPS-guided munitions in NATO's Serbia campaign; and up to the present, where Russia, the United States, and China are increasingly situating satellite capabilities at the heart of their command-and-control operations. All of this has also led to the development of antisatellite technologies. Concerns of a "Space Pearl Harbor" have been common within U.S. security circles, and, should a new global war begin with space powers on opposing sides, it seems likely that some of the opening shots will be in space.

Against this backdrop, it may seem difficult to separate the space exploration endeavors from the military space endeavors, given that they use much of the same technology, infrastructure, and personnel. To do so requires close examination of the specific histories of individual projects. For example, the development of liquid-fuel rocketry in the United States took place primarily within the military departments of the Department of Defense, including the army, the navy, and later the air force. It was clear that ballistic missiles were an important military capability that had emerged in the Second World War. The initial liquid-fuel rocketry experiments included launching seized V-2 rockets, often involving many of the same German engineers that had originally built, designed, and operated the V-2. Each of the three armed forces competed for the operational and developmental lead on ballistic missiles. Military interest therefore unquestionably drove the demand for liquid-fuel rocketry in its early stages. However, the development of boosters larger than what was needed for intercontinental ballistic missile (ICBM) uses was cut short. When General Medaris, commander of the Army Ballistic Missile Agency and von Braun's commander, argued that work on a larger booster should proceed, he failed to get approval, as "he could not cite a specific military need that would be served."[9] Both he and von Braun argued that "control" of outer space would ultimately require the more powerful boosters, but the broader DoD appropriating community remained unconvinced. The Saturn booster program did receive official DoD authorization in 1958 but even after much lobbying by Medaris and von Braun, it received only negligible funds.[10] In 1959, the DoD and NASA jointly endorsed a memorandum for President Eisenhower stating, "there is, at present, no clear military requirement for super boosters," while there was "a definite need for a super booster for civilian exploration purposes."[11] The military did not demand the level of liquid-fuel rocket development that was required for human missions to the Moon. The conclusion of the top military officials was that large boosters were to be pursued only if future weapons-systems requirements should evolve to a point where this could be justified militarily.

The Saturn program and the further development of the large liquid-fuel rockets required for the Apollo program were transferred to NASA. It is particularly notable that Secretary of Defense Robert McNamara, who lent his support to the pursuit of the Apollo program, testified that there was no "man-on-the-Moon requirement" from a military perspective.[12] There had been at least nominal cooperation between NASA and the DoD in the Gemini program, including the existence of a joint Gemini Program Planning Board "to ensure maximum attainment of objectives of value to both the NASA and the Department of Defense," although no significant DoD resources, other than the time of the personnel on the committee, were involved. There was no similar arrangement for Apollo, however, "because it [was] a mission that does not appear to have possible military ramifications."[13] Ultimately, the development of the Saturn V booster—the most powerful liquid-fuel rocket yet built and used for the single most expensive space exploration project to date—was not a response to military demand for the technology or capability. While McNamara recommended the pursuit of a manned lunar landing in a paper that he and NASA administrator James Webb presented jointly to President Kennedy, he did so specifically because he believed that "this nation needs to make a positive decision to pursue space projects aimed at enhancing national prestige."[14] Although McNamara argued that the Apollo program was important in the context of national security, this was not linked directly to military utility but rather to its "prestige"—its ability to signal American superiority over the Soviet Union. His support as secretary of defense derived from a recognition of the signaling value of the investment, which in the rhetoric of the 1960s was referred to as prestige.

Before considering further the impact of prestige, however, it is important to discuss briefly the fourth motivation identified in "Introduction to Outer Space," that of progress in scientific knowledge. The political role that science plays, and should play, in determining funding for space exploration has long been one of the most debated topics in spaceflight. Unlike with military needs, space exploration often directly contributes to scientific progress by uncovering new phenomena and acquiring new data on known phenomena. Furthermore, many of the central figures of space exploration and space policy are themselves scientists, making it difficult to isolate and identify the particular influences of scientific motives.

There are conflicting views expressed in the literature regarding the influence of scientific interests on space exploration policy. Dr. Lee A. DuBridge, president of the California Institute of Technology and a member of Eisenhower's PSAC, initially believed that "the predominating and overpowering reason for developing a substantial program of space exploration is the vast new extension of our knowledge which this will yield."[15] It appears that Eisenhower initially agreed, although both he and DuBridge would later concede the importance of prestige.[16] The secondary literature, however, has not lent much credence to the view that the government at the time conducted space exploration for scientific purposes. As Van Dyke put it, "no one who goes through the *Congressional Record* and the committee hearings and reports since Sputnik is likely to believe that this consideration [the expansion of human knowledge] has ever been uppermost in Congress."[17] Nonetheless, it was convenient to emphasize science as an altruistic motive for efforts that were designed primarily as impressive and symbolic signals.

The rhetoric of science ran high after the success of the Mercury program. Representative George Miller, chairman of the House Committee on Science and Astronautics from 1961 to 1973, proclaimed on the success of John Glenn's successful orbital flight that "This was in no way a stunt or an exhibition. It was . . . a great contribution to science. What we have seen today was a scientific experiment whose end object was not the entertainment of the people of this country and the world, although it means a lot to us in international prestige. That was not the end object either. Its end object is to wrest from space her secrets that can be used for the betterment of mankind."[18] John Glenn, however, undertook no scientific experiments and took no new scientific measurements while in orbit. Indeed, as Dr. Eberhardt Rechtin, at Caltech's Jet Propulsion Laboratory noted, "It is evident to most people, including most of the people in the Mercury program, that the 'pure science' in that program is zero."[19] Most writers of the political history of the space program have agreed on the limited role of science in setting the demand for human space exploration. Scientists in positions of budgetary and policy authority undeniably use the resources available to them to pursue scientific investigation and support the space science community—as had scientists in positions of authority at early American observatories. Indeed, the increasing size of the science budget within NASA over the past decades suggests that this trend has increased as political interest in the space program has declined. There is also a history of genuine, though varying, political interest in science for its own sake, either due to personal interest on the part of prominent politicians or due to the perceived downstream benefits of scientific research.[20] Eisenhower believed that the scientific merit of Explorer 1 and its link to the International Geophysical Year merited serious government support, although not absolute priority, for the initiative.[21] Science should thus indeed be considered one of the primary demands for space exploration, just as military and commercial applications are to be considered important motivations for space technology more generally. However, a demand for science was not the force that pushed forward the space race nor subsequent human spaceflight programs. For that, we must return to the third motive highlighted in "Introduction to Outer Space": prestige.

The importance of prestige in motivating space exploration was widely recognized by those observing the American space program in its early development. As but one example, Klaus Knorr, director of Princeton's Center of International Studies, argued in 1960 that "scientific and technological prestige will be a major objective motivating nations to participate in and, indeed, try to excel in space activities. . . . Space activities being especially glamorous, considerable advantages of international prestige are likely to accrue to the nation which assumes leadership in this enterprise."[22] The principal actors in the early American space program—

Eisenhower, Kennedy, McNamara, Johnson—also made frequent reference to prestige to describe the effects of and reasons for spaceflight. As a result, it is unsurprising that the concept has attained such a dominant explanatory role in the secondary literature. A review of the concept of prestige and its relevance to the space race will help explain its appeal.

The most focused investigation of the role of prestige in the history of the American space program is Giles Alston's 1990 Oxford dissertation "International Prestige and the American Space Programme."[23] Alston reviewed the academic literature on the phenomenon of prestige, drawing particular attention to the moral component implied in the concept. He notes how the difficulty of quantifying prestige has meant that it received little interest in the study of politics and international relations since the 1950s but that it had been a common subject of commentary and study in earlier periods. He reiterates the case that a competition for prestige was the dominant motivation for space activities during the space race—a view that was established in the earlier literature and that has continued since.

"Prestige," in its original usage, was at its root a type of illusion. The word had traditionally referred to a trick, such as the sleight of hand of a magician (as in "prestidigitation"), and its first modern usage in political terms was as a "dazzling influence," specifically in reference to Napoleon during his Hundred Days in 1815.[24] Drawing upon this etymological origin, F. S. Oliver noted that prestige "may be nothing more substantial than an effect produced upon the international imagination—in other words, an illusion . . . for the nation that possesses great prestige is thereby enabled to have its way, and to bring things to pass which it could never hope to achieve by its own forces."[25] E. H. Carr commented, "prestige means the recognition by other people of your strength. Prestige . . . is enormously important; for if your strength is recognized, you can generally achieve your aims without having to use it."[26] In an effort to determine the components of prestige, Shimbori and coauthors asked Japanese schoolchildren to rank countries and mention the first things that came to mind about them.[27] From the study, relative weights for different factors and their contributions to prestige were derived: total economic factors had a multiplier of 3; international politics, a multiplier of 3; military capability, a multiplier of 1; artistic and scientific matters, a multiplier of 4;

and international relations, a multiplier of 2. Although this result is certainly contingent upon the time (1961), place (Japan), and people (schoolchildren) involved, it suggests that achievements in science and art, both of which have reflections in space exploration, have the highest prestige multiplier. Space-related accomplishment clearly has all the attributes to achieve what Hans Morgenthau identified as the most important characteristic of international prestige, to "impress other nations with the power one's own nation actually possesses, or with the power it believes, or wants the other nations to believe, it possesses."[28]

These descriptions capture the key elements to national prestige. It is a trait of a nation that can be measured by the perception of others regarding its possession of certain attributes. It is strongly associated with power but also contains a moral element. Its possession affords different treatment by others as compared with those that do not possess it, and it is perception-based to the extent that its recognition by others is the critical metric. Prestige, therefore, is not an inherent characteristic of a nation so much as it is a perceived one.

The moral and the illusory elements of prestige have been closely linked in the literature on the application of the concept to the space race. The importance of the action being not just indicative of power but also, in its essence, "good" has been seen as a critical aspect of the political value of spaceflight activities. Lloyd Berkner, former chairman of the Space Science Board of the National Academy of Sciences, argued that a positive association with space exploration was nothing less than a result of human instinct: "Man prizes the idea of escape from the earth as the highest symbol of progress. Therefore, the nation that can capture and hold that symbol will carry the banner of world leadership. Consequently, leadership in space exploration has a real political meaning."[29] The nation-states that funded spaceflight projects were also quick to claim beneficent intentions for their actions—as emphasized in Nixon's goodwill tour of Asia connected with the success of Apollo 11 and most obviously in the plaque on the base of the Apollo 11 landing module stating, "We came in peace for all mankind." The Soviets made similar claims. Renowned African-American singer, civil rights activist, and communist sympathizer Paul Robeson was quoted in *Pravda* associating the flight of Yuri Gagarin and Vostok 1 with the contemporary struggles for independence in Africa: "Vostok flying over Africa is the light of the future freedom of the African continent, the source of faith and the strength of the millions of Africans fighting for their independence."[30] The propaganda machinery in the United States and the Soviet Union made effective use of space exploration as a symbol of goodwill.[31] The United States Information Agency (USIA) attempted to maximize the impact of U.S. successes with coordinated global campaigns. The ten-minute documentary "John Glenn Orbits the World" was distributed to 106 countries in thirty-two languages with an estimated international audience of over 200 million in the four months following the flight.[32] The Friendship 7 capsule piloted by John Glenn was exhibited in nearly thirty cities around the world, including Mexico City, Accra, London, Colombo, and Tokyo.[33] Two "Spacemobiles" with NASA-trained USIA officers toured Africa, Latin America, Europe, and India. Both sides made an active effort to imbue space exploration with positive meaning and moralistic associations around the world.

There is a problem, however, with this interpretation of the moral value of spaceflight and with the concept of prestige more generally. The acts of landing on the Moon or launching a man into space are not particularly credible signals of goodwill, peace, or solidarity. Although individuals and states can claim positive intentions and motivations, there is little inherent in the acts themselves that makes such claims believable. This is part of the illusion and perceptual sleight of hand inherent in the political value attached to spaceflight "prestige." The multibillion-dollar Apollo program, however, was not a grand illusion. While perception, specifically of global leadership, was indeed key to political support for spaceflight, an analysis in terms of prestige obfuscates the critical reason that spaceflight was able to foster this perception—the real, physical nature and technological complexity of the feats themselves. At its most basic level, the exploration of space is not a credible indication of positive moral intentions, nor is it sleight of hand; it is a complex and costly act that is a credible indicator of technical acumen, a capacity for brute force, and an abundance of resources and their centralized control. The credible transmission of this information—with all of the implications for geopolitical and economic power—is the root political value of spaceflight in the context of the Cold War. This is also the root of the concept of signaling. The political value of spaceflight is thus not its ability to effect the perception of global leadership in a vague, illusory manner through impressive and popular achievements. It is its ability to establish claims to technological and economic leadership through the credible signaling of these characteristics based on the difficult act of traveling into space.

Understanding the politics of the space race and American spaceflight through the wider lens of signaling, rather than prestige, allows for a new level of insight. As discussed in detail in the introduction, signaling is the process of a signaler credibly conveying information about itself to signal receivers. Good signals are costly to produce and thus can be credibly interpreted as indicating the possession of a vector of characteristics regarding the ability of the signaler to bear that cost; they are difficult to make and thus difficult to fake.[34] The implied signaling characteristic of space exploration through its significant resource requirements has been alluded to by many of the historical figures of space exploration. Explicit application of signaling theory to our understanding of space exploration, however, will help clarify the concept's explanatory value.

Let us examine a highly simplified example of space exploration's ability to signal at the national level. Let us assume that agents, in this case nations, are considering alliance decisions. There are many nations making this decision, but only two countries with which these nations can choose to ally themselves. All nations would like to ally themselves with the stronger of the two countries. Assume, however, that the agents exist in a world of asymmetric information and have only imperfect information about which nation is strongest. These nations would screen for the strongest ally by looking for the most credible (costly) signals that would convey a vector of desired characteristics. These characteristics might include abundant wealth, ability to muster resources effectively, technological mastery, ability to handle risk, willingness for peaceful cooperation and competition, and other characteristics sought after in alliance partners. Now assume that there is a potential signaling activity that is strongly correlated with many of the commonly desired traits and that this signal can be easily and objectively measured. Assume also that these two key states would like to have as many allies as possible, that the two states are not interested in cooperating, and that these other nations are continually updating their alliance preferences based on the signaling activities of the two states. It should not take a formal model to see that the result would be the two nations investing in this signaling activity while the nations of the world observe their performance and modify their behavior with respect to the two nations in response to the signals. The simplistic model reflects the simplistic way in which such decisions can sometimes be made.

Now let us further assume that each nation's actions are made up of some function of the weighted aggregate decisions of its individual citizens, a condition that also applies to the two key nation-states. The citizens of these two states also exist in a state of asymmetric information with regard to the qualities of their own nation and the qualities of the other leading nation. These citizens are also making alliance decisions of their own, primarily at the level of domestic politics. The costliness of the signal allows it to be interpreted by domestic citizens as a signal of a variety of domestic traits, including but not limited to competitiveness in foreign policy vis-à-vis the other state. It is also possible to credibly signal commitment to international cooperation or to industrial policies or a normative commitment to science, commerce, and other priorities, although the signal may be stronger (more closely correlated to the implied trait due to the type and magnitude of cost incurred) in some areas than in others. This signal is observed both by domestic and foreign observers over many sequences of decisions. This signal then, because of its high cost, visibility, and simple objective measurement, credibly transmits a vector of information about the signaling entity.

Space exploration can serve as such a signal because achievement is relatively objective, highly visible, and costly. Although the example has focused on the two-nation-states assumption, the signaling mechanism is scalable and generalizable for other scenarios as well. As Nobel Prize–winning economist Michael Spence points out, "the incentive to engage in activities that inform buyers is greatest for sellers with high quality products, but if they are successful, the incentive will trickle down through a spectrum of qualities."[35] The model is in fact scalable down to the level of the individual—the level for which signaling theory was initially developed—who is also looking to credibly signal his or her possession of traits (wealth, power, generosity, intellect) to broader society and to posterity, as could be seen in the signaling value attached to observatory investment by philanthropic patrons in search of recognition and legacy. This concept is a simplification of the complex set of political demands for space exploration, but it is a useful one. It provides a theoretical framework and a more nuanced context for understanding the political value of spaceflight and space exploration activities, and one that allows for a new perspective on the history of American spaceflight—from the bold nationalistic statement of the Apollo program to the cost-cutting compromise behind the creation of the International Space Station.

The political signaling value of space exploration was recognized well before the launch of Sputnik. In 1946, Project RAND, a think-tank formed within the Douglas Aircraft Company to provide policy analysis to the U.S. Armed Forces, published a report—its first ever—on the subject of a "Preliminary Design of an Experimental World-Circling Spaceship." In the introduction to the report, David Griggs, the first head of the RAND Physics Department and later chief scientist of the air force, identified the potentially powerful psychosociological impact that such a satellite would have: "the achievement of a satellite craft by the United States would inflame the imagination of mankind, and would probably produce repercussions in the world comparable to the explosion of the atomic bomb."[36] The RAND study explicitly recognized that the first satellite would be interpreted as a general signal of national strength and world leadership. In a follow-up paper nine months after the initial report, James Lipp, head of Project RAND's missile division, warned of the potentially dramatic consequences should the United States not be the first to achieve this feat: "Since mastery of the elements is a reliable index of material progress, the nation which first makes significant achievements in space travel will be acknowledged as the world leader in both military and scientific techniques. To visualize the impact on the world, one can imagine the consternation and admiration that would be felt here if the United States were to discover suddenly, that some other nation had already put up a successful satellite."[37] With RAND estimating a cost of $150 million in 1946—$11.9 billion dollars in 2015 GDP-ratio terms—the world-circling spaceship would indeed be a costly signal of technological capability.[38] Confronted with the report and its high cost-estimate, officials within the Department of Defense rejected its conclusions. Secretary of Defense

Erwin Wilson later characterized the satellite proposal as a useless stunt, saying that he would not be concerned if the Soviet Union achieved orbit first.[39]

In spite of this initial rebuff, the signaling value of the first spacecraft continued to be recognized within policy circles. In 1952, Aristid Grosse—physicist, president of the Temple University Research Institute in Philadelphia, and a member of Roosevelt's initial committee to investigate the atomic bomb—was spurred by von Braun's initial *Collier's* article to approach President Truman regarding the need to analyze the potential implications of satellite and space station development in the Soviet Union. The result of Grosse's initiative was a report that focused on the possibility of an unmanned satellite and concluded that a satellite launch "would be considered of utmost value by the members of the Soviet politburo" because of the signaling effect that such a satellite would have:

> The satellite would have the enormous advantage of influencing the minds of millions of people the world over during the so-called period of "cold war." . . . In the countries of Asia, where the star gazer since time immemorial has been influencing his countrymen, the spectacle of a man-made satellite would make a profound impression on the minds of the people. . . . Since the Soviet Union has been following us in the atomic and hydrogen bomb developments, it should not be excluded that the politburo might like to take the lead in the development of the satellite. . . . If the Soviet Union should accomplish this ahead of us it would be a serious blow to the technical and engineering prestige of America the world over. It would be used by Soviet propaganda for all it is worth.[40]

Importantly, these early reports focused specifically on the signaling value of space exploration and what that signal would mean in a broader strategic context, as well as on the direct military implications of satellites.

The satellite program made use of the investments in missile development, but a clear distinction was maintained between the signaling and scientific aspects of a satellite and the direct military value of a missile program. By 1953, it was becoming apparent that nuclear bombs could be made light enough to be transported effectively by rockets. Furthermore,

Department of Defense intelligence indicated that Soviet scientists were making significant progress with missile development. The air force assigned a ballistic missile program the highest priority, and the 1953 budget for the ICBM-IRBM program rose to over \$1 million (\$46 million as a 2015 GDP-ratio equivalent), and continued to increase thereafter.[41] Relatively speaking, this was a minor expenditure, comparable to the early stage of the V-2 rocket in Germany a decade earlier. When Eisenhower gave public authorization for satellite work in 1955, he stressed that a satellite would be "strictly for scientific purposes" and of no particular urgency. The responsibility for the satellite was given to the Vanguard project from the Naval Research Lab, primarily so that it would not interfere with the ballistic missile efforts at Wernher von Braun's army group, which were considered by far the higher priority.[42]

Eisenhower would later say that "our satellite program has never been conducted as a race with other nations."[43] At the time, this was more or less true. Von Braun and his missile team certainly saw themselves in competition with the Soviet engineers for the prestige of the first satellite launch. At the secretary level and above, however, the space-related potential of rockets was of secondary importance to their value as ballistic missiles. Similarly, the signaling value of satellites was initially considered of secondary importance to their value as tools for military reconnaissance, with Eisenhower prioritizing the development of the top-secret air force spy-satellite program, WS-117L, over Vanguard. The first scientific satellite was considered to be a worthy project for the nation, but it did not require a race. This is underlined by the bureaucratic lines that prohibited von Braun from using the successful test of his Jupiter C rocket in 1956 from achieving the first satellite launch. Although von Braun was aware of the important signal that would result if he had been allowed to orbit the first spacecraft, the administration was not convinced that there was enough benefit to warrant overturning the established boundaries on the satellite project. Administrative propriety was of sufficient concern that von Braun's group was ordered to ballast the inert fourth-stage motor of the Jupiter C with sand to ensure there was no "accidental" launch ahead of the Vanguard team. This was not because of arrogance that the Vanguard project would beat the Soviet Union, but simply due to the absence of a perceived urgent need, from either the president or the top military brass, to beat the Soviets in this endeavor.

There was as yet no demand for a large impressive signal vis-à-vis the Soviet Union—or at least no demand that it was believed a satellite could meet. This high-level lack of appreciation for the signaling value of a satellite was noted in the House Appropriations report on Project Vanguard written just after Sputnik: "The Vanguard Program was conceived in pre-Sputnik 1955 in an aura of unwarranted, but nonetheless real, national complacency concerning the technical supremacy of the United States."[44] This reaction to Sputnik helps underline the transmission of information that was part of the achievement. It was not simply a prestigious act that damaged American pride. It also convinced many Americans—through the evidence provided by a single, small object—that it was in effect the unveiling of a major challenge to the overall technical supremacy of the United States. It was this perceived loss of supremacy that would create a strong political demand for U.S. space exploration where previous arguments had initially failed.[45]

The launching of Sputnik in 1957 provided a strong and credible signal to the world in both the political arena and the electromagnetic spectrum. One of the important features of the satellite was the fact that it contained a radio transmitter whose signal could be picked up by amateur radio operators and listening stations across the globe, with the result that the world could independently verify what the Soviets had accomplished. The political potency of Sputnik was thus amplified by its immediate and widespread verifiability during a time when the conditions of the Cold War led to significant information asymmetries and difficulties in accessing reliable information. In 1957, information on income distribution, economic indicators, and technologies, even for one's own country, was relatively difficult to acquire. Information on social, political, and economic conditions was highly unreliable or sparse in many parts of the world and was particularly opaque across the borders of the Cold War. Statistical information on the Soviet Union was often nonexistent, and when it was present, it was difficult to trust. Moreover, books, radio, local newspapers, and word of mouth, which were the primary means of conveying information, could not easily traverse the political censorship and linguistic barriers between the Soviet Union and the West.

In contrast to the difficulty in obtaining reliable economic information across the boundaries of the Cold War, spacecraft were uniquely visible. A RAND / Rockefeller Foundation Conference in Washington in 1959 on "International Political Implications of Outer Space" stated that the unique visibility of space exploration made it of greater political importance than might otherwise have been the case. This visibility stemmed not only from the significant coverage of the event in the media or from the literal visibility of satellites—often it was the upper stage of the delivering launch vehicle that was seen by observers—it also stemmed from the global visibility of the heavens. Space is the only theater for which the entire world has essentially the same view. As such, the socioeconomic information encoded in Sputnik was more or less equally available to people at any major population center of the planet. In the pre-Internet era, no other signal had such widespread empirical verifiability. The psychological impact of the Sputnik moment was intense, as not only does the theater of space provide all viewers with equal visibility but it also positions them with the same perspective—looking up from the ground in wonder. The legendary, religious, and primal associations of the night sky served to further amplify the informational signal of Soviet technical power and mastery over the world.

Within this context of information asymmetry, the launch of Sputnik, verifiable simply by observing its flight across the night sky, was a credible signal to the entire world that the Soviet Union was the world leader in an advanced technology. It was undeniable that the launch of the world's first spaceship required significant intellectual and physical resources as well as a sophisticated command-and-control infrastructure. With limited information available on actual conditions in Russia, it also implied a more advanced economy, educational system, and, most importantly, military capacity than had previously been assumed. Literally overnight, with one signaling action, the global perception of the general ability of the Soviet Union increased substantially to the point where many now questioned the supremacy of the United States. If all one knows for certain about a nation is that it has been able to achieve the expensive and hitherto impossible act of launching an Earth-orbiting satellite, then that in itself is a credible signal that it is a very significant power. As Gabriel Almond, one of the fathers of modern comparative political science, observed in 1960, "In general, we may say that space competition has an importance for public opinion that is perhaps out of proportion to its over-all technological and military significance. That is to say, there is a tendency in popular opinion to take the space competition sector as indicative of over-all technological and military strength."[46] Because achievements in space were objective measures of social characteristics—at least in terms of technological ability, economic resources, and political will—they became a proxy measure for national standing at a time when other potential metrics were more difficult to measure.

The low-key response of the Eisenhower administration to Sputnik is consistent with an understanding of events based on signaling, since it demonstrates that the signal had the greatest impact for those who had the most significant information asymmetries regarding the Soviet Union. Eisenhower initially sought to publicly dismiss the significance of Sputnik, calling it just "one small ball in the air."[47] This was more than sheer bravado for the media. It also reflected Eisenhower's knowledge that his own plans were well advanced for a spy satellite to go beyond the U-2 planes. As commander in chief, Eisenhower had much better access to information and was thus able to judge the more fine grained reality of Soviet capabilities and to measure them in the broader context of the American economy. He underscored this in an article in *The Saturday Evening Post* after he left office. He suggested that other items be included in the "prestige race" with the Soviet Union, such as "our unique industrial accomplishments, our cars for almost everybody instead of just a favored few, our remarkable agricultural productivity, our supermarkets loaded with a profusion of appetizing foods."[48] Although all of these were more or less visible and objective measures of social achievement to Americans, few had firsthand or even reliable secondhand knowledge of these variables elsewhere in the world. The State Department's USIA wrote in an October 17, 1957, memo entitled "World Opinion and the Soviet Satellite: A Preliminary Evaluation" that "the technologically less advanced—the audience most impressed and dazzled by the Sputnik—are the audience most vulnerable to the attractions of the Soviet system. . . . The satellite, presented as the achievement and symbolic vindication of

the Soviet system, helps to lend credence to Soviet claims."[49] Space exploration achievements, unlike anecdotal reports on stocks of supermarket
goods, could be verified by radar systems and radio communication. In
the new, postcolonial states that were the battleground of the Cold War,
the paucity of available information on relative economic and social conditions in the United States and the Soviet Union made the Sputnik signal all the more powerful. Eisenhower did not seem to recognize the
central information transmission aspect of Sputnik, viewing it instead
within a more nebulous prestige context, and as a result he underappreciated the event's significance.

To evaluate the importance that foreign nations gave to Sputnik as a
signal of national ability, we can examine the information collected in the
1960s by Gabriel Almond, who had been head of the Enemy Information
Section of the Office of War Information during the Second World War.[50]
He discussed a number of surveys that were conducted in November 1957,
the month after the Sputnik launch. The surveys showed that 58 percent
of those in the United Kingdom believed that the Soviet Union was ahead
in scientific development, with only 20 percent believing the United
States to be ahead; the ratios were 49 percent:11 percent in France,
and 37 percent:23 percent in Italy.[51] Only in West Germany did the United
States maintain a perceived lead, with a ratio of 32 percent:36 percent. It
is tempting to speculate that proximity to Communist East Germany
(and more familiarity with Communist-sponsored science and technology)
may have influenced the German perception of U.S. dominance. Unfortunately, similar surveys were not conducted prior to Sputnik. We can,
however, confirm the impact of single-event space exploration signaling
by looking at the surveys conducted in October 1958 after the launch of
the first U.S. satellite Explorer 1. The change was dramatic. In the United
Kingdom, the net shift was from a 38 percent margin in favor of the Soviet
Union to a 13 percent margin in favor of the United States; in France, from
38 percent in favor of the Soviet Union to 14 percent in favor of the United
States; in Italy, from 14 percent in favor of the Soviet Union to 3 percent
in favor of the United States; and in West Germany, from 4 percent in
favor of the United States to 21 percent in favor of the United States.[52]

Alliance preferences in the four countries also seem to have been affected.
Another survey, conducted in May 1957 (before Sputnik), in November

1957, and in October 1958 asked, "At the present time, do you personally think that this country should be on the side of the West, on the side
of the East, or on neither side?" Over the course of the period, the results
were as follows: the change in the United Kingdom was a 4 percent net
decrease in siding with "the West" and a 9 percent increase toward neutrality; in France, a 14 percent increase in neutrality and a 15 percent decrease in siding with "the West"; in Italy, a 10 percent increase in
neutrality and a 7 percent increase in siding with "the East." Over this
period, there were obviously other contributing factors, and we must
be careful not to read too much into Almond's numbers. However, Almond's averaging of the changes in the surveys shows a number of interesting patterns: there is very little change in opinion favorable to the
United States; there is more change in opinion favorable to the Soviet
Union; and there is an increase in neutral opinions and a decrease in "no
opinions."[53] The signal seems to have changed perceptions significantly
but not drastically, supporting the Eisenhower administration's view that
the perception of the United States by European allies would not be
much affected.

The work of pioneering public opinion analyst Samuel Lubell also
suggested that there was no immediate panic in the American public
immediately after the Sputnik event.[54] Most of those surveyed dismissed
the satellite relatively easily and expressed sanguine opinions similar to
those of Eisenhower. Relatively few people repeated the criticisms found
in the newspapers. Rather than an immediate effect of the signal of
Sputnik, it was the subsequent campaign of public criticism by Senator
Lyndon B. Johnson that incited significant national apprehension about
space achievements. This campaign was given added fuel by the failure on
December 6, 1957, of the initial Vanguard launch, which had been intended
to match the Soviet Sputnik achievement but instead exploded on the
launching pad, with full television coverage for the American public. Concern about panic after the Vanguard failure was so great that the New
York Stock Exchange suspended trading within five minutes of the failure
at 11:50 A.M.[55] The media christened the event "Oopsnik," "Flopnik,"
and "Kaputnik."[56] The dramatic failure of Vanguard was read as a signal
that America lay exposed and could not muster the intellectual, financial,
and technical resources to match Soviet achievements.

Within this context, several key committees and boards within the Eisenhower administration clearly saw Sputnik as a blow to American prestige, linking it to the damage inflicted on America's international reputation by McCarthyism and the rise of the Soviet Bloc. Eisenhower himself, however, focused his public comments on the scientific and military aspects of the Soviet satellite, which he correctly judged to be modest.[57] Prestige, in Eisenhower's view, was not an important part of the equation. As Alston put it, "when the issue of prestige surfaced at all, it was usually as a signal for Eisenhower to explain . . . why money should not be spent on projects that presented their justification mainly in terms of prestige benefits."[58] Eisenhower, a war hero and the "liberator" of Europe, did not feel he needed to be concerned about illusory prestige and was instead worried that concerns over prestige were being generated in an attempt to extend the reach of what he would later call the military-industrial complex into the public treasury. Although this latter concern was legitimate, his dismissal of the psychological and perceptional importance of Sputnik meant that he left the door open for others to lay claim to the signaling value of spaceflight and to use it to their own political advantage.

Prominent members of Congress disagreed with Eisenhower and clearly saw space within a signaling competition context. Senator Lyndon Johnson was particularly critical of the Eisenhower administration's handling of the situation. Johnson mounted the first significant campaign of public criticism after Sputnik, focusing on the loss of status in the eyes of the world and on what Sputnik signaled about American complacency. Other American politicians also began to cite prestige and competition as reasons to ramp up the allocation of resources for the space program, despite Eisenhower's objections. As one member of the House Space Committee candidly asked NASA administrator T. Keith Glennan, "How much money would you need to get us on a program that would make us even with Russia . . . and probably leapfrog them . . . ? I want to be firstest with the mostest in space, and I just don't want to wait for years. How much money do we need to do it?"[59]

Technical officers of the U.S. government also recognized the signaling importance of space exploration. Asked to testify on the matter at a congressional hearing, George V. Allen, the director of the U.S. Information Agency, conceded, "No matter what we feel about it or how we may want it to be, we are in a space race with the Soviet Union. . . . Public opinion in the United States as well as overseas is going to put up what the Russians have done against what we have done. Every time the Russians do something, it is going to be marked up on a sort of chart. We are in a contest. There is no doubt about that, and no matter what we want to do about it, we are in this race."[60] Glennan, as Eisenhower's science-focused NASA administrator, was reluctant in his support of the prestige motive but ultimately agreed that competition with the Soviet Union was "the principal factor which determines the pace at which we pursue our goals. . . . It is clear to me that, as of the present, the enhancement of national prestige in a divided world has been and continues to be uppermost in the minds of the majority of people who have bothered to think about the matter of competition in the space arena. And this principally because of the fear that the loss of prestige that we have experienced for a time somehow upsets our equanimity and probably means, in some vague way, that we are second-best in *everything*."[61] It might seem that Glennan recognized that his job was to create the most clear and impressive signals possible. The signaling race was undeniable, and nonparticipation would send its own signals. For many of the key decision-makers in the U.S. government, it had become clear that this was not an option if they were to maintain the perception of global leadership.

Eventually, the combination of rhetoric and technological advances made even Eisenhower reluctantly recognize that the United States had to compete with the Soviet Union in the realm of space exploration. The Signal Communications by Orbiting Relay Equipment (SCORE) satellite became Eisenhower's own "space-coup," sending the first globally transmitted communication from space, a tape-recorded 1958 Christmas message conveying to the entire planet "America's wish for peace on earth and good will to men everywhere."[62] This, along with the impact of the Lunik II (which carried Soviet hammer-and-sickle flags to the Moon) and Lunik III (which orbited it), seems to have convinced Eisenhower of the signaling role of space activities. Soviet braggadocio over their space achievements also helped to spur the president into action. Khrushchev belittled the satellites that the United States had launched, calling them mere "oranges"; the first U.S. satellite Explorer weighed only 14 kilograms, and SCORE weighed in at 68 kilograms, compared with the 83.6-kilogram Sputnik, the 508-kilogram Sputnik II carrying Soviet space dog Laika, and the 390-kilogram Lunik series. Khrushchev's first official visit to the United States in 1959 happened to coincide with Lunik II's Moon voyage, which prompted a backhanded compliment. At a farm in Iowa where he enjoyed his first-ever hotdog, he quipped to the media, "We have beaten you to the moon, but you have beaten us in sausage-making."[63]

In the same year, when Eisenhower gave explicit spending goals to NASA administrator T. Keith Glennan, he divided the spending into three areas: first was to see to the success of the military program, third was for an orderly scientific program, but second was to see that real advances were achieved so that "the U.S. does not have to be ashamed no matter what other countries do; this is where the super-booster is needed."[64] Even Eisenhower, the small-government Republican, had accepted the importance of space visibility. The budget for space activities now began to increase significantly, from $331 million in 1959 to $964 million in 1961.[65] While Eisenhower had belatedly and begrudgingly come to recognize the space exploration signal, Kennedy would warmly embrace it.

John F. Kennedy had a clear vision of the signaling value of space-related accomplishment, as evidenced from his first comments on the subject prior to his election as president. In a speech in the Senate, Kennedy noted:

> If the Soviet Union was first in outer space, that is the most serious defeat the United States has suffered in many, many years. . . . Because we failed to recognize the impact that being first in outer space would have, the impression began to move around the world that the Soviet Union was on the march, that it had definite goals, that it knew how to accomplish them, that it was moving and that we were standing still. That is what we have to overcome, the psychological feeling in the world that the United States has reached maturity, that maybe our high noon has passed . . . and that now we are going into the long, slow afternoon.[66]

Kennedy focused specifically on the changes in international perception that Sputnik's signal had induced. His comments also make clear his more-encompassing global view as compared with Eisenhower's focus on Europe and the United States.

During Kennedy's years in Congress, he had shown an acute consciousness of the new states and new national leaders emerging through the decolonization process—states and leaders that would have no automatic tendency to adopt the economic and political system exemplified by the United States. Kennedy had made trips to India and Indochina in 1951, had gained attention by criticizing French policy in Algeria, had strongly advocated for foreign aid, and was a leading domestic figure on the subject of "third world nationalism."[67] He was alert to the battle for "hearts and minds" in these countries. In his speech in Congress on May 25, 1961, announcing the lunar program, he noted that these countries were "attempting to make a determination of which road they should take." He laid out his meaning explicitly: "I think the fact that the Soviet Union was ahead first in space in the Fifties had a tremendous impact upon a good many people who were attempting to make a determination as to whether they could meet their economic problems without engaging in a Marxist form of government."[68]

At the end of the 1950s, the United States appeared to be losing ground rapidly against a rising tide of Communism. The Soviet grip on Eastern Europe was ever tighter, with the Hungarian Revolution being crushed in 1956 and the border between East and West Berlin being effectively closed in 1957. Communist movements had begun to take hold in America's backyard of Latin America, with the CIA ousting a Communist government in Guatemala in 1954 and Castro seizing power in Cuba in 1959. The first U.S. troops were already being sent to Vietnam to train the South Vietnamese forces after the French were routed at Dien Bien Phu in 1954. The 1956 Suez Crisis, and the rise of new nationalist movements in Africa, had led to concerns about Soviet efforts to court influence and fill political power vacuums on the African continent.

New strategic thinking was also coming to the fore among the American policy elite, which placed a much higher premium on how the United States was perceived in the world. Throughout the Eisenhower years, the president and Secretary of State John Foster Dulles had espoused a military strategy based on "containment"—through NATO in Europe and a set of global alliances with anti-Soviet governments—along with a capacity for "massive retaliation" or "massive response." In the years just prior to Kennedy's presidency, however, the thinking had begun to shift. In 1957, the same year as Sputnik was launched, Henry Kissinger had published his first major book, *Nuclear Weapons and Foreign Policy*.[69] The best-selling book was an expansion on an earlier article he had written in *Foreign Affairs*, entitled "Military Policy and the Defense of the 'Grey Areas.'"[70] He set out to address the dilemma of having to choose between nuclear Armageddon and "defeat without war," and in doing so he elaborated the concept of flexible response. This included contemplating the possibility of limited nuclear war, but he also argued for a much wider arsenal of usable force and influence, including what has later come to be known as "soft power." This shift in thinking meant that the issue of prestige and global reputation was increasingly perceived as having concrete strategic significance. This perception was further heightened by the emergence in 1961 of the so-called Non-Aligned Movement—founded by the leaders of Yugoslavia, India, Egypt, Indonesia, and Ghana—a group of countries where influence would have to be earned on a continuing basis as opposed to being taken for granted based on a predisposition toward the United States.

The status of America's global reputation became one of the key issues of the presidential campaign in 1960 and an important point of differentiation between Kennedy and Nixon. Kennedy put strong relations with the new states in Asia, and a robust economy for the purpose of signaling the success of capitalism, at the forefront of his campaign. The link between America's reputation and the new states in Asia was particularly prominent in the second and third televised presidential debates.[71] Importantly, however, neither Kennedy nor his advisors drew a connection during the election campaign between a robust space program and impressing the Third World. Although public USIA reports had concluded that Soviet space achievements were creating a perception of Soviet technological leadership and eroding the U.S. image, this issue did not emerge significantly in the election. In the record of campaign speeches and appearances between August 1 and November 7, 1960, Kennedy made reference to space 54 times, 33 of which were as part of broader statements about being second, while prestige itself (unrelated to space) was mentioned more than 250 times.[72] Even with Johnson, the space program champion, on the ticket as vice presidential candidate, space was apparently not considered a promising election issue. This was at least in part due to Nixon's use of the American space program record to rebut Kennedy's own points about America's declining prestige, commenting, "In the spirit of Halloween, Mr. Kennedy promises you a treat, but in the end its just a trick. . . . This one is the claim that Soviet Russia is first and we are second in space exploration. Mr. Kennedy is wrong. The facts are that we have successfully launched 26 earth satellites, and 2 space probes, 13 satellites are still in orbit, 8 of which are still transmitting information. The Russians have launched only 6 satellites and 2 space probes and only 1 satellite is still in orbit and it is not transmitting."[73] Indeed, Nixon displayed both the more accurate information and the more enterprising position on space during the election. He laid out his policy in a speech in Cincinnati that proposed the development of nuclear-thermal propulsion, a space station in 1966–1967, circumlunar flights in 1966–1968, and human lunar landings in the early 1970s. He promised to be "second to no one in the long stride into space."[74] By contrast, Kennedy made no specific commitment to space policy goals, only to reversing the decline in American's global reputation. In laying enormous stress on reputation as an overarching foreign policy goal, however, it is not surprising that Kennedy would take the decision to pursue the Apollo program in order to signal the reestablishment of U.S. leadership.

It was the flight of Russian cosmonaut Yuri Gagarin, the first man of the Space Age, that induced Kennedy to demand a mission that would one-up the Soviets. NASA administrator James Webb had previously tried to get approval for the Apollo program in March 1960 and had been rebuffed. But Gagarin's flight on April 12, 1961, led to a new media outcry, as *The New York Times* declared Gagarin's flight new evidence of "Soviet superiority" while the Red Square celebrations were carried globally by the BBC.[75] Less than a week after Gagarin's feat came the humiliating Bay of Pigs invasion on April 17–19—the failure of which constituted another powerful signal, albeit an unintended one. Even before that, Kennedy had ordered a full-scale inquiry into the space program and had decided that some response was needed, although most sources agree that he had yet to make a final decision on his course of action.[76] After the Bay of Pigs, he sent Vice President Johnson a memo asking:

> 1. Do we have a chance to beat the Soviets by putting a laboratory in space, or by a trip around the moon, or by a rocket to land on the moon, or by a rocket to go to the moon and back with a man? Is there any other space program which promises dramatic results in which we could win?
> 2. How much additional would it cost?
> 3. Are we working 24 hours a day on existing programs? If not, why not? If not, will you make recommendations to me as to how work can be speeded up?[77]

The second and third questions are process related, but the first addresses the root issue: Is there a way to win? The perception of competition is taken for granted. The question is, What space achievement can the nation undertake that would let the world know that America is superior to the Soviet Union? Kennedy had clearly come to appreciate the general signaling value of space exploration and was determined to make use of that characteristic as a means of repositioning America's global reputation.

Johnson's consultations and responses indicated his agreement with this position, noting, "This country should be realistic and recognize that other nations, regardless of their appreciation of our idealistic values, will tend to align themselves with the country which they believe will be the world leader—the winner in the long run. Dramatic accomplishments in space are being increasingly identified as a major indicator of world leadership."[78] The political and signaling nature of the decision to go to the Moon is further underscored by the omission of the President's Science Advisory Committee from Johnson's consulting process. When it came time for the joint submission of a response plan by James Webb of NASA and Department of Defense Secretary Robert McNamara, competition for global leadership was portrayed as the prime motivator of space exploration: "*This nation needs to make a positive decision to pursue space projects aimed at enhancing national prestige.* Our attainments are a major element in the international competition between the Soviet system and our own"[79] (emphasis in original). As Kennedy saw it, NASA's goal was "the preservation of the role of the United States as a leader," and "to demonstrate to a watching world that it is first in the field of technology and science."[80]

Kennedy, McNamara, and Webb all perceived the way in which space exploration acted as a signal of national abilities and character. Kennedy's Rice University speech in September 1962, in which he announced his intention to pursue the Apollo program, evoked intrinsic values, calling the exploration of space "one of the great adventures of all time." It emphasized the signaling aspect of space activities, stating that a country must engage in space exploration endeavors "because they are hard, because that goal will serve to organize and measure the best of our energies and skills."[81] Landing on the Moon, Kennedy saw, would be an effective signal of American supremacy because it would be an objective measure of national skills and, above all, because it was hard and costly—the root of an effective signal. In his address to Congress on the matter, he further noted the potential "impact of this adventure on the minds of men everywhere, who are attempting to make a determination of which road they should take."[82] He also explicitly made the connection between the power of a signal and its cost: "no single space project in this period will be more impressive to mankind . . . and none will be so difficult or expensive to accomplish."[83] The signaling effect was the fundamental basis for Kennedy's support for the Apollo program, as is clear when he proclaimed that "if we are to go only halfway, or reduce our sights in the face of difficulty, in my judgment it would be better not to go at all."[84] Here Kennedy portrayed the Apollo program as a winner-take-all competition, not something that could be justified on its own intrinsic merits, but something that could only be justified in the context of a competitive imperative. Kennedy had clearly understood the holistic signaling character of space exploration, as had James Webb: "In the minds of millions, dramatic space achievements have become today's symbol of tomorrow's scientific and technical supremacy. There is, without a doubt, a tendency to equate space and the future."[85] Robert McNamara spelled out his own understanding in even more detail: "All large scale space programs require the mobilization of resources on a national scale. They require the development and successful application of the most advanced technologies. Dramatic achievements in space, therefore, symbolize the technological power and organizing capacity of a nation. It is for reasons such as these that major achievements in space contribute to national prestige. This is true even though the scientific, commercial or military value of the undertaking may, by ordinary standards, be marginal or economically unjustified."[86] McNamara understood the process through which space-related achievements signaled national power and had even come to see that the signal was the only valid political justification for undertaking the enterprise. For all the major decision-makers who enabled the Apollo program, the resources were seen as dedicated to national prestige—recognition of the signaling function of space exploration.

Although prestige might be a sufficient explanatory concept for the early history of the decision to go to the Moon, Kennedy's offer to Khrushchev of a cooperative lunar landing has complicated narratives that put prestige at the center. Within a month of his speech to Congress announcing his intention for a manned Moon mission, Kennedy suggested the possibility of a joint lunar expedition to Khrushchev. This occurred at the June 4, 1961, Vienna Summit, where Kennedy pursued the idea with Khrushchev over lunch but received an ambiguous response and no commitment.[87] Khrushchev later warmed to the idea after the orbital flight of John Glenn, however, and suggested cooperation in his congratulatory note. Kennedy and the State Department worked to develop concrete proposals, including two tandem weather satellites, tracking stations, and satellite communications. The result would be brief discussions between technical representatives on both sides, but enthusiasm for the cooperation faded due to general disagreement on disarmament. Although Kennedy reiterated the idea as late as September 1963 in a speech to the United Nations, his assassination two months later put an end to the prospect. The abortive attempt to create a cooperative lunar program thus left no major programmatic legacy, but it did leave a historical challenge to those seeking to explain the events of the space race solely on the basis of competition for prestige.

One rejoinder is that, regardless of Kennedy's intentions, the relatively broad base of support for the Apollo program suffered erosion when a cooperative venture was proposed—thus suggesting that a competitive dynamic was indeed the driving force that sustained the American space program. Kennedy's apparent change of course for Apollo created a backlash that drained political support and is likely to have played a part in the cuts by the Manned Space Flight Subcommittee of $259 million for fiscal year (FY) 1964, $120 million directly from Apollo, plus a further reduction of $90 million on the House floor. In the heated discussion of the FY 1964 budget, a Republican amendment to the bill was added that prohibited the use of any funds for a joint human lunar landing with any Communist nation. The political demand for space exploration as a competitive signaling opportunity clearly outweighed any inclination toward cooperation.

A signaling interpretation makes it possible, however, to incorporate Kennedy's offer of cooperation into the same framework as the competition for prestige. Because of the substantial resources required for a lunar landing, Kennedy's willingness to contemplate sharing the massive project with the Soviet Union was a way of providing a strong signal of openness toward substantive and broad-based overall cooperation. Indeed, Kennedy's overture on the Apollo program was the first explicit offer of a program of large-scale cooperation between the two superpowers since the Second World War. Instead of the costliness of space exploration being used to signal superpower leadership over a perceived rival, Kennedy was using it as a credible signal of a general willingness to cooperate. The fact that he repeated the offer at the United Nations in 1963, in spite of congressional concerns, is of particular significance. In the wake of the 1962 Cuban Missile Crisis, when conflict between the superpowers had brought the world to the brink of nuclear war, Kennedy understood the value of seeking to redirect the signaling power of the space program toward building trust and confidence.

The power of space exploration to signal cooperative commitment was recognized in the United States as early as the legislation that created NASA—the National Aeronautics and Space Act of 1958. The House Select Committee reporting on the legislation saw international cooperation at the heart of the new agency: "It is necessary to make real, aggressive efforts at forming international programs, at developing the purely international frame of mind in which lies this earth's only ultimate stability."[88] The committee further declared that the new space agency "must be organized first as an active agent of international cooperation and ultimately as the basis for an international organization."[89] Senator Lyndon Johnson alluded poetically to the way in which cooperation through space programs might lead to peace: "Men who have worked together to reach the stars are not likely to descend together into the depths of war and desolation."[90] A study by the staff of the House Space Committee in 1961 directly recognized the mechanisms through which resource expenditure on space exploration could contribute to peaceful international relations, noting that "the absorption of energies, resources, imagination, and aggressiveness in pursuit of the space adventure may become an effective way of maintaining peace."[91] At the root of this theory was a recognition that mutual expenditure of resources on an expensive and high-visibility project would send a credible signal of cooperation and would bind the behavior of both parties toward joint endeavor.

The intent to signal cooperation was at the fore in the first post-Apollo spaceflight program—the 1975 Apollo-Soyuz Test Project (ASTP). ASTP featured no fundamentally new operational or technological knowledge nor any significant gain in military knowledge for either side. It was, however, at the time the most resource-intensive project of cooperation between the two sides and was part of the culmination of the détente process. Although the mission itself did not happen until 1975, its origins can be dated back to an exchange of letters in 1962 between Kennedy and Khrushchev after the Vienna meeting. With both agreeing in principle to increased cooperation, Kennedy asked his administration to prepare "new and concrete proposals for immediate projects of common action."[92] NASA, the White House, and the State Department prepared a list of space projects that could serve as the first such action. This led to the first discussions between NASA Deputy Director Hugh Dryden and Anatoli Blagonravov, former president of the Academy of Artillery Sciences and then Soviet representative to the United Nations Committee on the Peaceful Uses of Outer Space. These negotiations led to the first concrete projects of cooperation—in data exchange and satellite tracking. These projects were important in the buildup to détente and in providing the groundwork for ASTP. Nixon and Brezhnev signed the agreement that included ASTP in 1972 as part of a number of cooperative undertakings meant to signal political commitment to détente.[93]

The official U.S. budget for ASTP was around $300 million ($3.2 billion in 2015 GDP-ratio terms), although that did not fully include the cost of the Apollo Command Module, Apollo Service Module, and Saturn 1B rocket, which had already been constructed for the Apollo lunar program.[94] Of this expenditure, only $10 million was for scientific work. As Olin Teague, chair of the House Space Committee acknowledged, it was "strictly a political, psychological effort."[95] That ASTP's significant expenditure was justified purely on its signaling benefit is widely accepted; in fact, this reinforced its potency as a signal with respect to Soviet-U.S. cooperation.

Cooperative efforts in space were also used as visible signals of alliance building with existing allies by both sides. By the end of 1962, the United States had made arrangements for cooperation in space activities with sixty-one countries, with a specific focus on sounding-rocket development in countries ranging from Canada to Pakistan.[96] The Soviet Union signed cooperative agreements on space activities with France in 1966, and the Interkosmos program saw the participation of cosmonauts from thirteen Soviet allies, from Afghanistan and Bulgaria to Syria and Vietnam. The United States responded with collaborative initiatives on the Space Shuttle program with Canada and West Germany, and invited flights with nationals from Mexico and Saudi Arabia. Although the inclusion of foreign nationals on already-scheduled flights had a relatively low marginal cost to the Soviet Union and the United States, they were nonetheless high-value tokens of favor for the partner countries who could not sustain the cost of developing their own spaceflight capabilities. Veblen noted the way in which this type of cooperation is also part of signaling leadership and superiority: "As wealth accumulates on his hands, his own unaided effort will not avail to sufficiently put his opulence in evidence by this method. The aid of friends and competitors is therefore brought in by resorting to the giving of valuable presents and expensive feasts and entertainments."[97] Looking at these actions through a signaling framework allows the separate motivations of "prestige" and "peace"—of competition and cooperation—to be understood as functioning through the same mechanism.

While providing a unifying context for space exploration policies aimed at both peace and prestige, signaling theory also provides new insight into other important events in American space history. This includes understanding how concerns over negative signaling contributed to the decision by the Nixon administration to develop the space shuttle. Negative signaling is an aversionlike phenomenon that, as Veblen explained, arises out of the arms race of conspicuous consumption: "it is much more difficult to recede from a scale of expenditure once adopted than it is to extend the accustomed scale in response to an accession of wealth."[98] This results in a lock-in effect, whereby an "element of the standard of living which set out with being primarily wasteful, ends with becoming, in the apprehension of the consumer, a necessary of life."[99] Or, as it is more colloquially put—a luxury, once acquired, becomes a necessity. Once a signal has been established, it becomes difficult, even politically impossible, for the signaler to cease the signaling activity, as it would cause a significant loss of reputation. Within this context, there is significant evidence to suggest that such a loss aversion to negative signaling has been an important element in the resource allocations for spaceflight in the post-Apollo period.

Nixon's decision to develop the space shuttle is a good example of the role that concerns over negative signaling have played in the American space program. Beginning in 1969, NASA began to search for a new program to follow Apollo. The agency initially proposed to develop a series of expensive elements for a space exploration architecture that would eventually lead to a landing on Mars. The Nixon administration, however, had read the public mood, which was against additional expensive spaceflight extravaganzas, and vetoed NASA's initial proposals until the agency was left with a single element of this elaborate architecture—the space shuttle.[100] It was a close fight for even this new expenditure.[101] The negotiation and principal budgetary discussion for the space shuttle took place between NASA and the Office of Management and Budget (OMB) and was not an initiative of the president in the way that Apollo had been for Kennedy. That such a major space program was established first within the broader government bureaucracy is itself a sign of the acceptance that large space exploration expenditures had achieved within government budgets. The argument that ultimately carried the day, however, was one that highlighted the signaling value of spaceflight—and more specifically the negative signal that the failure to pursue a new spaceflight project would send.

When NASA administrator James Fletcher made his case for the space shuttle, his first point was a simple statement, that "the U.S. cannot forego [sic] manned space flight because for the U.S. not to be in space, while others do have men in space, is unthinkable, and a position which Americans cannot accept."[102] Caspar Weinberger, deputy director of the OMB, who had significant purview over NASA's budget and was generally hostile to new NASA requests, echoed Fletcher's statement. Weinberger was very blunt about the signal that would be sent if the United States failed to commit to a new spaceflight program: "It would be confirming in some respects, a belief that I fear is gaining credence at home and abroad: that our best years are behind us, that we are turning inward, reducing our defense commitments, and voluntarily starting to give up our super-power status, and our desire to maintain world superiority."[103] Interestingly, Weinberger did not have a strong opinion as to *what* the major new spaceflight program should be—space shuttles and nuclear rockets were just two options that seemed acceptable to him—only that there should be one. On the memo, Nixon wrote "OK" and "I agree with Cap" to indicate that he supported Weinberger's analysis.[104]

President Nixon's close confidant and counsel John Ehrlichman similarly suggested that it was a concern over a loss of reputation, rather than the potential for a gain, that shaped Nixon's view of the space shuttle: "We had to be at the leading edge of this kind of applied technological development. And if we weren't, then a great deal of national virtue was lost, and our standing in the world and all that."[105] When the space shuttle program was ultimately put to Nixon after extensive negotiations between NASA, White House staff, and the OMB, Nixon asked if it was a good investment but quickly added, "even if it was not a good investment, the nation would have to do it anyways, because space flight was here to stay. Men are flying in space now and will continue to fly in space, and we'd best be part of it."[106] Whether or not it was wasteful, space exploration had become an element of the conspicuous investment of a superpower and a cost that the U.S. government had accepted as now being a necessity of political life.

Further evidence for the signaling value of spaceflight can be seen in the origins of the American Space Station Freedom program under Reagan and its evolution into the International Space Station program under the Clinton administration.[107] Reagan's announcement of his space station initiative was unique in that it had an explicitly electoral context. He first announced his decision in his 1984 State of the Union address, where the space station was the only major new initiative proposed. With 1984 an election year, the incumbent president was in a difficult position. As a conservative facing a large government deficit, he had to appear to be restraining government spending. At the same time, he needed to back up his claim that, under his leadership, the nation had a fresh new start and that it was "morning in America" again. His speech addressed these themes and identified what he believed the initiative would signal:

> Our second great goal is to build on America's pioneer spirit. . . . A sparkling economy spurs initiatives, sunrise industries, and makes older ones more competitive.
>
> Nowhere is this more important than our next frontier: space. Nowhere do we so effectively demonstrate our technological leadership and ability to make life better on Earth. . . . Opportunities and jobs will multiply as we cross new thresholds of knowledge and reach deeper into the unknown.
>
> Our progress in space—taking giant steps for all mankind—is a tribute to American teamwork and excellence. Our finest minds in government, industry, and academia have all pulled together. And we can be proud to say: We are first; we are the best; and we are so because we're free.
>
> America has always been greatest when we dared to be great. We can reach for greatness again. We can follow our dreams to distant stars, living and working in space for peaceful, economic, and scientific gain. Tonight, I am directing NASA to develop a permanently manned space station and to do it within a decade.[108]

This high rhetoric made for a bold statement of national renewal, and Reagan's investment in the new space station could be seen as a credible commitment of his dedication to those ideals. It was also about the only new commitment that he could afford with a budget deficit of roughly $185 billion for the 1985 fiscal year.[109] For Reagan, the space station, projected initially at a cost of only $8 billion, was thus a high-visibility program, replete with future promises at a relatively marginal cost. It was this ability to seemingly signal a bright future for the nation at a comparatively low cost that made it such an attractive element for Reagan's 1984 election campaign.

The space station initiative is the only NASA program to have been integrated into a presidential campaign platform at such a high level. In the Reagan-Bush 1984 official campaign brochure, the space station featured prominently. The brochure noted that "Americans were ready to make a new beginning. So we elected President Ronald Reagan and Vice President George Bush to lead us into a more promising future" and that the "President has challenged us to move forward again, to unite behind four great goals to keep America free, secure and at peace for the '80s."[110] The first goal was to "Ensure steady non-inflationary economic growth," the third to "Strengthen our traditional values," but the second was to "Develop space, America's next frontier."[111] The brochure further declared that in order to achieve this goal, "President Reagan has proposed the construction of a permanent manned space station."[112] President Reagan also made a giant model of the space station a centerpiece of the May 1984 London G-7 Economic Summit, where he invited leaders of Europe, Canada, and Japan to join in the project. Reagan maintained his emphasis on the station throughout his campaign, including at a major rally in Fairfield, Connecticut, less than two weeks from election day, where he waxed lyrical on the wonders of the space shuttle and stated that "we've committed America to meet a great challenge—to build a permanently manned space station before this decade is out."[113] He finished his speech with an appeal to a farsighted vision: "America is never going to give up its special mission on this Earth—never. There are new worlds on the horizon, and we're not going to stop until we all get there together."[114] The space station, with its significant but not overwhelming cost, could be portrayed as a credible indication that Reagan intended to fulfill this view, and as a result it became an important part of his successful election platform. The signaling value of spaceflight had again contributed to a major new initiative. It would also, however, lead to that program being scaled back when a new president wished to signal fiscal responsibility and international cooperation rather than the dawning of a new day for the nation.

President Clinton's conversion of Reagan's Space Station Freedom project into a reduced-budget cooperative venture with Russia is an example of the negative effect that spaceflight's signaling value can also have on spaceflight expenditures. In the years between Reagan's announcement and when the Clinton administration took office in January 1993, NASA had spent $11.2 billion on the project with few pieces of hardware completed and none of them yet launched into space.[115] Whereas Reagan had used the space station to signal his commitment to a bright future for the nation, Clinton had other signals that he wanted to send: fiscal responsibility in the face of difficult economic conditions, and a willingness to cooperate with Russia following the collapse of the Soviet Union. In Clinton's 1993 State of the Union address, he emphasized core issues of domestic politics, bemoaning "two decades of low productivity and stagnant wages; persistent unemployment and underemployment; years of huge government deficits."[116] His only reference to space was to use its distance to emphasize the scale of the debt problem: "I well remember, twelve years ago Ronald Reagan stood at this podium and told the American people that if our debt were stacked in dollar bills, the stack would reach sixty-seven miles into space. Today, that stack would reach two hundred and sixty-seven miles."[117] With his focus on the debt, he stated his intention to reduce government spending by $246 billion over five years. Exactly four months after his 1993 State of the Union speech, Clinton would make it clear that the space station would be an important part of those cuts.

In his "Statement on the Space Station Program" of June 17, 1993, Clinton began on an ominous note for NASA: "At a time when our long-term economic strength depends on our technological leadership and our ability to reduce the deficit, we must invest in technology but invest wisely, making the best possible use of every dollar."[118] Although there remained the usual references to space as being vital to "technological leadership," the overriding note was one of retrenchment rather than advance: "I am calling for the U.S. to work with our international partners to develop a reduced-cost, scaled-down version of the original Space Station Freedom. At the same time, I will also seek to enhance and expand the opportunities for international participation in the space station project so that the space station can serve as a model of nations coming together in peaceful cooperation."[119] Although the move signaled a credible commitment to cooperate with Russia—as Kennedy's overtures to Khrushchev might have—it just as pointedly signaled a commitment to tackling the debt while at the same time continuing to invest in American leadership in technology. Clinton claimed that with "deep cuts in future development and operations costs," the redesigned program would save more than $4 billion over five years, and more than $18 billion over the two-decade life of the program.[120] These space station "savings" represented almost 2 percent of the total Clinton cuts. Although NASA's budget represented only 1 percent of total federal expenditures at the time, spaceflight's visibility had made it a target for deeper cuts to signal the seriousness of the Clinton administration's attack on the debt. As in the other major American space program decisions—Apollo, the space shuttle, Space Station Freedom—the decision to pursue the International Space Station had signaling as a major political motivation. The International Space Station, however, is a unique reminder that, although spaceflight has largely been used as a signal for American leadership, there is also a reverse side to the coin. Reductions in spaceflight funding can also be used to signal a commitment to austerity and the intent to refocus on a more "down-to-earth" agenda.

While an identification of the overall signaling function of spaceflight clarifies its exchange value in a political context, it also raises a number of questions and problems. The most fundamental of these is the effectiveness of the signals. If there have been political desires and decisions to pursue space exploration for its signaling value, to what extent have these signals been effective? This is an inherently difficult question to answer, and to address it comprehensively would require opinion surveys of the world population before and after individual space exploration achievements. We only have a few of these types of surveys—such as those by the USIA and Almond, referenced earlier—and the results are ambiguous.[121] The effect of the Sputnik signal was strong but heterogeneous: those geographically closer to, and presumably with more detailed knowledge of, the Soviet Union were less influenced by the signal than those farther away. Similarly, although a significant portion of the world population was impressed by the success of the Apollo program, it seems equally clear that a number of important actors were not. General Võ Nguyên Giáp and the North Vietnamese, for example, were hardly overwhelmed by American spaceflight achievements and continued to press their attacks while their enemy walked on the Moon: their understanding of the real power of the United States, at least in a military sense, was at most only marginally influenced by space signals, and they ultimately prevailed on the battlefield. Although the political desire for signaling may have been a driving force, the effectiveness of the space signal was ambiguous even at the height of the Space Age, and it became even more so in the subsequent decades.

There is also a case to be made that space signals can be as deceptive as they are informative. The cost of a spaceflight achievement means it can be an informative signal relative to other sources of information—but at the same time, because spaceflight is not a perfect representation of the characteristics it signals, it can also mislead and create false impressions. By its nature, a signal is only a proxy for an underlying set of characteristics that remain largely unknown to observers. Savvy actors can thus intentionally manipulate perceptions using space signals, and there is evidence to suggest that this has been the case—such as with Khrushchev's desire to leverage the Soviet space program to project an international image of a powerful and robust Soviet economy, which was at odds with domestic reality. An American example of this intentionally deceptive use of space signaling can be seen in President Reagan's Strategic Defense Initiative (SDI). According to some SDI scholars and participants, the program was conceived of, at least in part, as an expensive signal that would severely strain the Soviet Union's industrial base if it tried to match it, and which would sow dissension and undermine Soviet confidence if it didn't.[122] From 1983, the DoD and the Reagan administration avidly promoted SDI to the public, even though many experts argued that the system had major technical obstacles that precluded realistic implementation. Even with some $26 billion spent on the program in the 1980s and early 1990s, an operational space-based intercept system was never fielded.[123] Despite never having been implemented, however, there is evidence to suggest that the program had an effect: at least some high-ranking decision-makers in the Soviet Union seem to have believed SDI to be a genuine signal of American technical and economic superiority. According to some accounts, this had a demoralizing effect on Soviet leadership and played a role in the Soviet collapse. It is also worth noting that the conceptual and advocacy origins of SDI have been traced to members of the space colonization advocacy group the L5 Society, making it another example of a program promoted by private individuals who offered a powerful space signal to politicians in exchange for resources that they hoped to use to satisfy their own intrinsic preferences for spaceflight.[124]

The potential use of spaceflight signaling for intentional deception also raises the question of whether some of these signaling investments were also unintentional instruments of self-deception. To what extent did space achievements lead to overconfidence among American and Soviet leadership in the Cold War? To what extent did these achievements encourage damaging overextensions, such as with the Soviet invasion of Afghanistan and the extended American campaign in Vietnam? To what extent have historic space achievements been given undue weight as relevant signals of national strength in the subsequent decades, long after the underlying conditions that produced the initial signals have changed? To what extent did the success of the Apollo program help to foster an overly optimistic perception of the organizational capabilities of NASA in the decades after its achievement?[125] These are complex questions deserving further analysis and research; they cannot be answered here, but they are important to consider when thinking about the overall public utility of the signaling value of space exploration activities.

Signaling theory provides a useful analytical tool for understanding the context of resource allocation decisions made during the space race and the political value of spaceflight. It enables the consolidation of a number of the commonly identified motivations for spaceflight programs into a simple framework with explanatory power that applies over a broad range of space history. The demands for prestige, pride, peace, competition, and cooperation, which have been identified as among the most basic drivers of space exploration expenditure, can all be reasonably approximated by a demand for signaling. This simplified model is not meant to account for all of the complex demands and motivations behind space-related resource allocation. It does, however, provide a new interpretive framework within which to place key political decisions.

Signaling considerations have played a role in a wide range of endeavors throughout the historical continuum of American space exploration. As we have seen earlier with the construction of early American observatories, signaling can be a powerful motive at multiple levels—individuals, institutions, civic communities, and national governments. Amid the competitive antagonism of the Cold War, the importance of the space exploration signal reached new heights. The efforts of the early pioneers of liquid-fuel rocketry, spurred on by wartime support from the military, had reached a point of technological breakthrough that opened up the potential for spectacular new space exploration milestones to be achieved. The milestones were within the grasp of two world powers that were locked in a political struggle for global influence and dominance. Moreover, that struggle was one of conflicting economic and political systems, where the signaling characteristic of space exploration gave it high value as a means of projecting system superiority.

Space exploration could be used to signal many things, but above all, it credibly signaled an ability to muster resources, organize capacity, harness technology, and surmount challenge—in short, the ability to possess and project power. The more dramatic the accomplishments and the more resources and technological sophistication required to achieve them, the more effective the signal became: a nation that could send a man to the Moon was seen as one that could accomplish the seemingly impossible. The strength of the signal was further amplified by the transition in communication technologies. The mid-twentieth century's information asymmetries, particularly across the Cold War divide, meant that space exploration achievements were effectively a type of globally verifiable information that came to be seen as a proxy measure for overall economic development, technological capacity, and military strength. At the same time, the beginning of the rapid proliferation of television provided a means of amplifying the signal by its rapid distribution to both domestic and worldwide audiences. As hundreds of millions of people across the planet watched Neil Armstrong plant an American flag into the lunar soil—though the television picture may have been fuzzy—the signal of American power was clear.

The principal improvement of the signaling framework is that its emphasis of a *characteristic* of space exploration—rather than a *rationale for* space exploration, as is more common in the literature—allows for the production of space exploration to be understood as part of an economic exchange. Although at a geopolitical level it was the American government that "produced" the achievements of the space race, at the project level of actual spacecraft and missions, it was the spaceflight engineers, scientists, and advocates who were the producers of items of political value that were, in effect, traded to the American government in exchange for financial support for themselves and their projects. A signaling perspective thus takes account of both the intrinsic interests driving the technical pioneers and dedicated promoters on the supply side of the exchange, as well as the demand-side requirement for a quid pro quo on the part of those allocating the resources. When Korolev's team succeeded in convincing Khrushchev to allow Sputnik to be launched, it was not an attempt by the politicians to create a compelling symbol of the new Soviet society. The Sputnik project itself was largely the result of the personal efforts of Sergei Korolev, Mikhail Tikhonravov, and other Soviet designers, leveraging military demands and the infrastructure and technology that those demands allowed them to build.[126] The act of launching Sputnik nonetheless had the characteristic of signaling the abilities of the Soviet Union to the world—as well as establishing the powerful signaling characteristic of spaceflight. With the political value of space exploration thus established, the producers of spaceflight signals—in America and the Soviet Union—now had a basis for large-scale exchange with government patrons. A view of spaceflight as having a fundamental political value in its signaling ability thus does not mean that signaling is the motivation of spaceflight per se. Rather, the signaling characteristic of spaceflight is simply the exchange value of a product that, for its actual producers (the engineers, scientists, and managers of spaceflight), is often intrinsically motivated. This provides a view of space history that allows the intrinsic motivations of the individual virtuoso engineers and scientists to be recognized as paramount in the development of spaceflight while at the same time identifying the signaling value of their products as being their principal exchange good in the context of political funding.

The idea that signaling is the basis of trade between the intrinsically motivated supply side of the spaceflight production equation and the politically motivated demand side provides a basic model—at least for government-funded space exploration—that, though simple, is still more instructive than the "prestige thesis" that has often been dominant in the explanation of the space race. William Sims Bainbridge's *The Spaceflight Revolution: A Sociological Study,* which details the myriad techniques that the engineer-entrepreneurs and other early spaceflight advocates used to acquire the resources for their projects, notes how "the spacemen" manipulated politicians' concerns over "the nebulous heavens of international prestige" and were able to "sell false solutions," since, as he puts it, America did not seem to have benefited from successfully landing on the Moon.[127] Though there is a great deal of insightful observation and argument in his work, it brings to mind the original etymology of "prestige" as a sleight of hand, as does Gerard DeGroot in *The Dark Side of the Moon* when he claims that the Apollo program was a "brilliant deception" and a "glorious swindle."[128] Although there is certainly some truth to Bainbridge's opening statement, that it was "not the public will but private fanaticism that drove men to the moon," it is a mistake to ignore or deny the political value that space exploration had in the context of the Cold War.[129] If we leave behind the concept of prestige, however, and build on a framework that recognizes the signaling characteristic of spaceflight, it becomes possible to recognize both the driving importance of the intrinsic interests of the spaceflight pioneers and advocates, and the political value that was obtained by allocating resources toward a strong spaceflight-related signal. Viewed from this perspective, the Cold War American space exploration program was the result of a grand bargain that bridged the supply and demand sides of the spaceflight production equation at an unprecedented scale.

In understanding the space race and the lunar landings as part of a signaling process, it is also important to recognize the unique conditions that made that initial grand bargain possible. It was a rare conjuncture of technological events and geopolitical circumstances that allowed space exploration activities to develop at such a rapid pace in the middle of the twentieth century. The space race was also enabled by an expansionary postwar global economy, with especially strong growth in American GDP in the 1960s, which made the massive expenditures required to produce the costly spaceflight signals both affordable and politically salable. Space-exploration-related expenditures in the 1960s could thus be viewed more enthusiastically than would be the case during the oil price shocks and stagflation of the 1970s or the anxious concern over the U.S. deficit in the 1990s. Not only was the Apollo program at the crest of the momentum of the first generation of spaceflight pioneers, it came at a point when their activities had immense perceived and real political value, allowing for a large-scale exchange of resources for signaling goods between politicians and the spaceflight industry. While spaceflight accomplishments continue to have real signaling value, the heightened political demands that drove the extraordinary growth of the American space industry in the 1950s and 1960s have dissipated and may not coalesce again in the near future.

From this perspective, the Apollo program should not be seen as the classic model of American space exploration, but rather as an anomaly. From a long-run historical perspective, this "Apollo Anomaly" represented an exciting new paradigm for American space exploration, but ultimately a short-lived and ephemeral one. NASA and the American spaceflight community have continued to try to emulate the superficial conditions of the anomaly—with new presidential directions and planetary destinations—in the hope that the anomalous funding, appetite for risk, and political momentum will return, but to no avail. It is not due to a lack of direction or leadership that American astronauts have not left low-Earth orbit in over forty years. In a world with easier access to information, to the point of ubiquitous information overload, the signaling value of space exploration has diminished. NASA has correspondingly received a fairly stable and substantial budget over the past few decades—the national luxury of spaceflight, having been acquired, remains a political necessity—but nothing comparable to the peak that it experienced during the Apollo program.

There may, however, be some benefits to the decreased signaling value of spaceflight. If the nature of a strong signal is in part that it is costly to produce, then it has at least in part been the expense of spaceflight that has thus far provided much of its enduring political value. Yet many of those men and women who have, throughout the century or so since Goddard's first experiments, contributed to the development of spaceflight have themselves desired to travel into space and have not: the cost has thus far been prohibitive for the vast majority of them, and the incentives have not effectively aligned within the government to substantially reduce the cost of spaceflight. As the signaling value of spaceflight decreases, it becomes more apparent to intrinsically motivated spaceflight engineers and entrepreneurs that the future of space exploration—and their personal potential to experience it—under wholly governmental auspices may be limited, and they may begin to seek out and cultivate other patrons. As happened in the early twentieth century for Goddard and in the nineteenth century with astronomical observatories, some of these efforts have been successful, and patrons in the private sector have once again begun to resume some of their traditional historical role in the funding and pursuit of American space exploration. In the conclusion we will consider what the combination of the different phenomena explored in this book—signaling, intrinsic motivations, and the long shifting history of public and private funding for space exploration—means for the future.

THE NEXT SPACE PATRONS

Mankind will not remain forever confined to the Earth. In pursuit of light and space it will, timidly at first, probe the limits of the atmosphere and later extend its control to the entire solar system.

—*Konstantin Tsiolkovsky,* Letter to B. N. Vorobyev, *1911*

What do we learn from this long-run perspective on American space exploration? How does it change our understanding of the history of spaceflight? How does it change our understanding of the present? This book has provided an economic perspective on two centuries of history, with examinations of early American observatories, the rocket development program of Robert Goddard, and the political history of the space race. Although the subjects covered have been wide-ranging, together they present a new view of American space history, one that challenges the dominant narrative of space exploration as an inherently governmental activity. From them a new narrative emerges, that of the Long Space Age, a narrative that in the *longue durée* reveals personal initiative to have been the wellspring of American space exploration activities for over a century before the Space Age and that shows signaling-based political support to be dramatic but short-lived in comparison with the long history of private funding.

Together, these studies argue for a reorientation of the overall economic narrative of American space exploration. By focusing more attention on the extensive history of activities in the pre–Space Age era it becomes clear that private funding and intrinsic motivations have played a much greater role than have generally been considered—comparable at times to space exploration's considerable political value as a signaling device. Though its political signaling value has been the principal source of demand for the grandest spaceflight efforts, it is also a value that is temporary and fleeting when considered in historical perspective. The more persistent long-run forces are those of individual passions and intrinsic motivations, in private pursuit of the dream of a future in space. In its full historical context, American space exploration is more often a private initiative than a governmental one.

Examining the history of the Long Space Age as a whole presents three important and related insights. Perhaps the most unexpected discovery has been the empirical evidence concerning the long-run importance of private funding in the financing of American space exploration—the precedent of privately funded hundred-million-dollar and even billion-dollar resource-share-equivalent space exploration projects in the nineteenth and early twentieth centuries. This shows that the twenty-first-century trend of wealthy individuals, such as Paul Allen, Jeff Bezos, and Elon Musk, devoting some of their resources to the exploration of space is not an emerging one. Rather it is a persistent, enduring trend that is now *re*emerging. These men are space entrepreneurs and patrons in the long tradition of James Lick, George Ellery Hale, Andrew Carnegie, and Harry Guggenheim. After an important but comparatively brief period of geopolitical space theater during the Cold War, these new explorers of the heavens are placing their bets, dedicating their resources, and picking up where their predecessors left off. This is the heart of the Long Space Age—that in the long historical perspective, the American movement out into space is much more than the story of "one giant leap" by its government in service of geopolitical competition; it is a cumulative story of the many small steps of its people, some taken with the support of their government, but many of the most important supported by private resources and individual will alone.

The expenditure data collected for astronomical observatories and Goddard's programs, along with NASA's programmatic history, also suggest that there has been significant volatility in long-run space exploration expenditures, often with a sharp expenditure peak, followed by a decline. The histograms of total U.S. observatory expenditures revealed this pattern, with peaks in the 1840s, 1870s, and 1920s, followed by decades of relative decline. These peaks correspond to the culmination of the American Observatory Movement, the founding of the Lick Observatory, and the founding of the Palomar Observatory, respectively. On a smaller scale, Goddard experienced a number of revenue peaks during the course of his program, in 1918, during the First World War, when working for the U.S. Army Signal Corps; in 1931, with the first grant from Daniel Guggenheim; and in 1942, during the Second World War, when working for the navy's Bureau of Aeronautics. NASA also experienced a significant funding peak during the Apollo program, reaching its maximum height of expenditure in 1965, which was subsequently followed by a sharp decline in the 1970s. Together, the data suggest that the time evolution of American space exploration expenditures has included significant variation for almost two centuries, and that continued peaks and troughs in expenditures should be expected. It is worth noting that the peaks identified were the result of roughly three types of activity: an elite social movement, individual large private-sector investments, and geopolitical conflict.

Despite the repeated peaks and troughs of expenditures, however, the persistence of private-sector and intrinsic support throughout the Long Space Age also suggest that the cultural momentum of spaceflight is far from expended. Although some of the political momentum may have ebbed, when single individuals, either through financial or intellectual resources, are able to direct historical development, cultural influences and personal motivations can become more important than geopolitics or financial incentives. The driving forces of spaceflight can thus be seen as a set of personal urges and passions. As Arthur C. Clarke put it, "Any 'reasons' we may give for wanting to cross space are afterthoughts, excuses tacked on because we feel we ought, rationally, to have them. They are true but superfluous, except for the practical value they may have when we try to enlist the support of those who may not share our particular enthusiasm for astronautics, yet can appreciate the benefits which it may bring, and the repercussions these will have upon the causes for which they, too, feel deeply."[1] This personal, intrinsic value gives spaceflight activities an enduring robustness. As long as there are individuals who desire to fulfill the vision of a space-faring future for themselves and for humanity, and who possess the requisite resources, talent, and willpower, the Long Space Age is far from over.

In addition to these insights into the historical patterns of American space exploration expenditures, there are a few lessons learned that are relevant for current issues in space policy. One is that the regular creation of new institutions has been an important feature of American space history. The history of American astronomy is in part a history of the new astronomical institutions that were established and that supported researchers with funds and facilities; the Harvard College Observatory, the Lick Observatory, the Mount Wilson Observatory, and the Palomar Observatory are four examples that remain active and important institutions of American astronomy today. Although John Quincy Adams failed in his initial attempt to create a national observatory, his efforts led to the creation of the Smithsonian, which became the first federal institution to invest in spaceflight technology development when it supported Goddard's research in 1917. In this context, NASA appears not as the sole defining institution of American space exploration, but rather as one institution among many—one with a changing role and identity as the political environment, U.S. aerospace industrial base, and private-sector ambitions and capabilities change around it. It is far from inconceivable to envision that NASA, as a federal institution, might experience further institutional evolution. NASA may have started out as an organization responsible for the development and operation of its own launch vehicles and spacecraft, but perhaps in the future it may focus more on its robotic and human space exploration missions while transitioning launch-vehicle development and operation to the private sector—particularly if the relevant private-sector entities continue to build on the lead they have recently developed over the government in the area of reusable launch systems, which have the potential to significantly reduce the cost of spaceflight. It is worth noting that intrinsically motivated individuals who desire to personally explore and travel through space tend to be very interested in reducing the cost of spaceflight so that they can do more of it.

The long-run history of American space exploration should teach us to see these transitions—the rise of new institutions and the transformation of old ones—as part of the natural course of American space development.

The American space community would do well to accept this historical fact and perhaps spend less time and energy worrying about how to preserve and rehabilitate old institutions and more effort on encouraging the emergence of new institutions of American space exploration that could provide new impetus amid the current circumstances and opportunities. It is worth noting that a large fraction of the institutions that have had the longest-lived influence—the Lick Observatory, the Mount Wilson Observatory, the Smithsonian—have been independently established with ample endowments each provided by a single wealthy philanthropist who had the objective of creating a permanent legacy.

American space exploration has been the product of a multiplicity of people and projects over decades and centuries that have evolved over time into a network of public and private interests and institutions. At times, private interests and institutions have had the strongest influence within the network, such as during the Observatory Movement or the private funding of Robert Goddard, and at other times it has been public interests and institutions that have been strongest, such as during the Cold War or Goddard's military funding. Understanding space exploration as the product of networks of public and private actors highlights the ability of administrations and individuals to create new nodes and dynamics within these networks in order to advance objectives in space. As Bromberg has pointed out in *NASA and the Space Industry,* NASA itself is a consummate example of a public-private innovation network, with its missions produced by a combination of government civil servants and private-sector workers.[2] New billionaire-backed private-sector space companies and projects thus do not reinvent space exploration out of whole cloth, but rather they add new nodes to an existing network that has been evolving for over a hundred years. Their value then is not necessarily that they are more efficient or innovative than governmental efforts. Rather, it is that they increase the diversity and complexity of the overall network of production and therefore contribute to its strength, by, for example, counteracting trends toward consolidation in the aerospace industry, by creating a multiplicity of human spaceflight organizational cultures, and by resisting the tendency within governmental engineering and scientific-research organizations, such as NASA, to primarily back their own projects. Furthermore, the creation of new private institutions of space exploration can, under certain circumstances, be considered endogenous variables. History shows that billionaire interest in space exploration is something that can be cultivated. Hale cultivated Carnegie's interest, and Goddard and Lindbergh cultivated Guggenheim's. There is much to learn about the practical techniques of motivating grand new projects of space exploration in the correspondence of these early space entrepreneurs, with their extensive networks of patrons and friends, and the long list of private-sector projects that have been encouraged in this way suggests that, at least in part, it will be through similar such relationships and efforts that future space exploration projects will emerge.

Although the discussion here has been confined almost exclusively to the national context of the United States, understanding the role that signaling has played in American space exploration expenditures is also helpful for understanding its role within an international context. While the signaling value of space achievements has been reduced for the United States relative to their peak value in the Cold War, it still remains strong—as evidenced by the national and international reaction to the landing of the Curiosity rover on Mars in the summer of 2012. It also remains strong for rising nations—in particular China, with its stronger conditions of relative information asymmetry across its population—that wish to signal that they have general economic and technological capabilities competitive with those of the United States. As each national space program, and the international spaceflight community as a whole, considers what exploration objectives to set and what partnerships to engage in to achieve them, planners and policy makers should consider these questions with the geopolitical and domestic signaling implications of each decision in mind. While a variety of political motives drive national space policies, an appreciation for the often overriding role that the signaling characteristic of space exploration has had in motivating expenditures can aid in defining programs and projects that find political purchase.

Small satellites and emerging commercial spaceflight capabilities also now have the potential to offer space achievements at significant reductions in cost. Although under conditions of perfect information this would correspond to a commensurate reduction in their signaling value, in practice these achievements still retain much of their value due to imperfect information and general association with their more expensive predecessors and counterparts. Low-cost satellites and space probes thus present a cost-effective signaling opportunity for nations wishing to announce and demonstrate their capabilities. Similarly, we can expect that there will be at least some international demand for commercial human spaceflight capabilities, should they emerge: seven individuals have already paid tens of millions of dollars each to fly into space on the Russian Soyuz vehicle, and geopolitical conspicuous consumption alone may be enough to motivate some nations to purchase, rather than develop, their own space vehicles and maintain their own astronauts, provided that the costs are not prohibitive. The international demand for national signaling of technological and economic capability can be expected to continue to contribute to the global development of space exploration, and, indeed, national policies should consider both international space agencies and foreign private-sector space efforts as nodes that can interact with their own national space innovation networks.

I hope this book also provides historical perspective on the question of who should pay for space exploration. It has been established that the private sector has paid for American space exploration for the majority of its history, although it was public-sector investments that realized the emergence of human spaceflight in the Cold War. We are now at a moment at which the private sector has reasserted some of its historic role, making foundational investments in the generation of new spaceflight capabilities and commercial companies. At the same time, it has been government support—in terms of technical expertise, tests facilities, and programmatic funding—that has allowed the most advanced of these capabilities to progress as rapidly as they have. Private-sector funding for space exploration has also been shown to be far from an unalloyed good. While there are important examples of genuine philanthropy in the history of astronomy, there are also examples of individuals who explicitly looked to use the positive publicity associated with space exploration projects to try to ameliorate some justly deserved negative reputations. Private-sector projects were also subject to many of the same budgetary and schedule overruns that have become familiar in modern government space programs. While some private-sector space exploration institutions contributed greatly to the expansion of our store of knowledge about the universe, others were bedeviled by problems and challenges of execution that greatly limited their contributions and effectiveness. Although I have argued that we must recognize the important and leading role that private-sector funding and initiative has had in the long history of American space exploration, that history also makes clear that neither the public sector nor the private sector has a monopoly on either technological innovation or on misguided efforts.

How then should we balance the increasing role and initiative of private-sector American spaceflight capabilities with the responsibility of government investment to promote the public good? This is the critical question of American space policy for the early twenty-first century. From a public-policy perspective, it's not only a question of how much the nation should give a federal agency like NASA to explore and develop space, but also how much the nation should provide in terms of support to American citizens and organizations that wish to explore and develop space. How much is it worth, collectively, to the United States of America, for some of its citizens to return to the Moon again, to fly by Venus, to orbit Mars, or, indeed, to develop settlements on Mars? It may not always be worth the amount that it would cost NASA to develop and operate such missions, but it may be worth supporting private endeavors, at some fraction of the total cost, in the form of public-private partnerships—much as John Jacob Astor IV had envisioned in 1894. Although such support might well be provided through NASA, asking the question in this way removes the focus from an individual institution of American space exploration and places the attention on the development of American spaceflight and space-development capabilities more broadly. The economic perspective presented here also suggests that we should recognize that the core desires for spaceflight and space exploration are intrinsic, personal motivations. This places the onus for further spaceflight development and space exploration firmly on those parts of the American citizenry who possess and share these interests. It is—in short—up to those with the resources, will, and capability to explore space to organize themselves and their materials so that they can do so. It is then up to the rest of the population, through the American political process and governmental administration, as well as through voluntary expenditure and contribution, to figure out how much to support these individuals and organizations—whether they are situated within the government or in the private sector—based on the amount of public and private good that is expected to result from their achievements.

Although this analysis has highlighted the long-run importance of independent groups of Americans and private-sector funding in the economic narrative of American space exploration, it has also shown that the financial scale on which these groups have operated is far smaller than the scale at which government spaceflight programs have been funded. Though independent groups of Americans are increasingly channeling the long-run forces of American space history to advance their spaceflight projects, the public funds that have propelled government-led space exploration programs are well over an order of magnitude larger. In many respects, Ormsby MacKnight Mitchel has been proven right when he declared in the 1840s that "under a republican form of government, the people must hold, with respect, to all great scientific enterprises, that position of patrons which in monarchial governments is held by Kings and Emperors."[3] In the United States, despite the rising importance of private-sector individuals in spaceflight funding, it is still in fact the American people—through taxes and public institutions such as NASA— that remain by far the largest patrons of space exploration. This might seem like a paradox at the heart of the American history of space exploration: it is the efforts and resources of individuals and independent groups that have the longest history in American space exploration, and yet it is the federal government that has provided the resources and political motivation for the largest and most-recent projects. There thus seem to be two conflicting trends governing the development of space exploration in this country: a long-run, moderate private-expenditure trend, and a short-run, major government-expenditure trend. I argue that the moderate resources of independent groups of Americans have been the leading force over the long history of American space exploration; but I also recognize the strength of the antithesis, that it has been government funding and institutions that have provided the principal economic impetus to the largest and most-recent space exploration projects.

The strength of both the thesis and the antithesis, however, suggests the potential strength of a spaceflight program that achieves a synthesis of these two historical forces. Might there be a space policy that could synthesize these forces and provide renewed momentum for the nation's efforts to explore the solar system? Examining the long-run history of American space exploration suggests that an effective synthesis of these forces might emerge in the form of a national strategy that coordinates the efforts of NASA and the intrinsically motivated individuals that are funding and driving the most effective private-sector space exploration projects. If private funders, individual virtuosos, and independent groups of Americans have been the long-run historical drivers of American space exploration, it would be wise for government policy to, in some instances, focus on the areas of private-sector initiatives and expertise that have garnered the most support in terms of independent financing and use the unrivaled resources and funding of government agencies to amplify and accelerate those independent efforts that are deemed to have technical merit and to be in the national interest. In other instances, it would be wise for government policy to ensure that those areas where public-sector institutions have retained the most relevant technical expertise, and in areas that require decades-long commitments and consistent funding unlikely to be provided for by the private sector, remain robust and capable nodes in the network of public and private entities that make American space exploration possible. Policy makers and space program executives should be mindful both that nonprofit, private-sector funding for space exploration has been an important source of major projects of space exploration in the United States, but also that it was only when the American people became the most generous patrons of space exploration in history through their taxes that Americans first began to fly into space and to land on other worlds. Rather than suggest that there is a particular optimal policy path to be pursued with regard to the balance between public and private institutions, the history of the Long Space Age should serve as a reminder that the social and economic circumstances that have supported American space exploration are contextual, complex, and changing.

Above all, the Long Space Age should serve as a reminder to maintain a long historical perspective when considering such epochal events as the extension of humanity into the solar system. There is an understandable sense of urgency in the efforts of modern space exploration advocates, fanned by the fear of losing the institutional momentum provided by the Apollo program, by the changeable winds of national politics, and by the earnest personal desires to be part of the exodus from the cradle of

Earth. And yet, the long-run history of space exploration shows it to be an endeavor more robust in its support than often thought. The history of the Long Space Age is thus also an encouragement to be patient and to prepare for the long run rather than the short sprint. The long tradition of American space exploration is a reminder that spaceflight stands today on roots that are centuries old. We would be wise to embrace these roots and to focus on strategies for the development and exploration of the solar system that are suited to the fundamental drivers of our movement out into the cosmos and that have in them the seeds for further growth and evolution over centuries and millennia.

NOTES

Introduction

1. *The Economist,* "The End of the Space Age," cover article, June 30, 2011; S. J. Dick and R. Launius, eds., *Critical Issues in Spaceflight* (Washington, DC, 2006).

2. Bonestell, C., and Ley, W., *The Conquest of Space* (New York, 1950), p. 95.

3. Zahavi, A., and Zahavi, A., *The Handicap Principle: A Missing Piece of Darwin's Puzzle* (New York, 1999); Zahavi, A., "Mate Selection: A Selection for a Handicap," *Journal of Theoretical Biology* 53 (1975), pp. 205–14; Grafen, A., "Biological Signals as Handicaps," *Journal of Theoretical Biology* 144 (1990), pp. 517–546; Johnstone, R. A., "Sexual Selection, Honest Advertisement and the Handicap Principle: Reviewing the Evidence," *Biological Reviews* 70 (1995), pp. 1–65.

4. Veblen, T., *The Theory of the Leisure Class: An Economic Study of Institutions* (New York, 1899), p. 68.

5. Ibid., p. 86.

6. Offer, A., "Lecture Notes on Economic and Social History," Michaelmas term, Oxford University, 2007.

7. Malone, T. W., and Lepper, M. R., "Making Learning Fun: A Taxonomic Model of Intrinsic Motivations for Learning," in *Aptitude, Learning, and Instruction: III. Conative and Affective Process Analysis,* ed. R. E. Snow and M. J. Farr (Hillsdale, NJ, 1987), pp. 223–253.

8. Maslow, A. H., "A Theory of Human Motivation," *Psychological Review* 50, no. 4 (1943), pp. 370–396; Maslow, A. H., *Motivation and Personality* (New York, 1954).

9. U.S. Congress, Comm. on Aeronautical and Space Sciences, *Scientists' Testimony on Space Goals,* Hearings, 88th Cong., 1st sess. (1963), p. 51.

1. Piety, Pioneers, and Patriots

1. A good general resource for discussion and calculation of relative value can be found online at www.measuringworth.com, which was used to calculate the GDP-ratio and PWC-ratio equivalent values in this paper in November 2016.

2. Krisciunas, K., *Astronomical Centers of the World* (Cambridge, 1988).

3. The source references for the costs can be found in the text descriptions of each project. The then-year costs of the projects are represented with as much accuracy as can be determined from the sources used. The 2015 GDP-ratio equivalent values have been rounded to two significant digits. The base year for the calculation of the 2015 GDP-ratio equivalent value is made from the year of commitment of funds, except in the cases where the specifics of funding are known over a number of years or in observatories with long development times, in which case the equivalent values of each year's costs are used and summed as outlined in the reference or a midpoint in the observatories' development is chosen. Ideally, the individual annual expenditures on the projects would be determined, the 2015 GDP-ratio equivalent value of those expenditures calculated and then summed over the entire program. Annual expenditure records for these projects, however, are not readily available, and significant further research will be required to achieve these more-accurate estimates (if the data is even available to achieve them). The current estimates, do, however, present an appropriate order of magnitude.

4. Johnston, L. D., and Williamson, S. H., "Sources and Techniques Used in Construction of Annual GDP, 1790–Present," MeasuringWorth, 2008, https://www.measuringworth.com/uscompare/GDPsource06.htm; Officer, L., "Characteristics of the Production-Worker Compensation Series," https://www.measuringworth.com/datasets/uswage/pwcsessay.php.

5. McCray, P., *Giant Telescope: Astronomical Ambition and the Promise of Technology* (Cambridge, 2004), p. 3.

6. NASA, Cost Analysis Data Requirements (CADRe) Database, Office of Program Analysis and Evaluation.

7. John Hopkins University Applied Physics Laboratory, "Near Earth Asteroid Rendezvous—Frequently Asked Questions," 1998–2000, http://near.jhuapl.edu/intro/faq.html.

8. Miller, H., *Dollars for Research: Science and Its Patrons in Nineteenth-Century America* (Seattle, 1970); Kealey, T., *Sex, Science and Profits: How People Evolved to Make Money* (London, 2008); Kealey, T., *The Economic Laws of Scientific Research* (London, 1996).

9. Yeomans, D., "The Origins of North American Astronomy—Seventeenth Century," *ISIS* 68, no. 1 (1977), p. 415.

10. Tolles, F., "Philadelphia's First Scientist James Logan," *ISIS* 47, no. 1 (1956), p. 22.

11. Greene, J., "Some Aspects of American Astronomy, 1750–1815," *ISIS* 45, no. 4 (1954), p. 340.

12. Fithian, P. V., *Journal & Letters of Philip Vickers Fithian, 1773–1774: A Plantation Tutor of the Old Dominion,* ed. H. D. Farish, Williamsburg Restoration Historical Studies, no. 3 (Williamsburg, 1943), p. 117.

13. Greene, "Some Aspects of American Astronomy, p. 339.

14. Bell, W., "Astronomical Observatories of the American Philosophical Society, 1769–1843," *Proceedings of the American Philosophical Society* 108, no. 1 (1964), p. 7.

15. Loomis, E., "Astronomical Observatories in the United States," *Harper's New Monthly Magazine* 13 (June 1856), p. 26.

16. Hindle, B., *The Pursuit of Science in Revolutionary America* (Chapel Hill, 1956), p. 165.

17. Musto, D., "A Survey of the American Observatory Movement," *Vistas in Astronomy* 9, no. 1 (1968), p. 88.

18. Hindle, *The Pursuit of Science in Revolutionary America*, p. 170.

19. Bernard Cohen, I., *Some Early Tools of American Science* (Cambridge, 1950), p. 157.

20. Jefferson, T., "Notes on Virginia," in *The Writings of Thomas Jefferson*, vol. 2, ed. A. E. Bergh (Washington, DC, 1907), p. 95.

21. Hindle, *The Pursuit of Science in Revolutionary America*, p. 337.

22. Bell, "Astronomical Observatories of the American Philosophical Society," p. 8.

23. Paullin, C., "Early Movements for a National Observatory, 1802–1842," Records of the Columbia Historical Society, Washington, DC, vol. 25 (1923), pp. 40–43.

24. Bell, "Astronomical Observatories of the American Philosophical Society," p. 11.

25. Jefferson, T., and Cabell, J., *Early History of the University of Virginia* (Richmond, 1856), p. 189. GDP-ratio equivalent and unskilled-labor-wage-ratio equivalent values in this chapter have been calculated using measuringworth.com: Johnston and Williamson, "Sources and Techniques Used in Construction of Annual GDP," and Williamson, S. H., "Sources for the Unskilled Wage," MeasuringWorth (2009), https://www.measuringworth.com/datasets/uswage/source.php; and Margo, R., *Wages and Labor Markets in the United States, 1820–1860* (Chicago, 2000), p. 117.

26. Musto, "A Survey of the American Observatory Movement," p. 89.

27. United States, President, *The Addresses and Messages of Presidents of the United States, from 1789 to 1839* (New York, 1839), p. 299.

28. Musto, "A Survey of the American Observatory Movement," p. 89. It is an interesting side note that the project costs were proposed to be offset by a sale of information for a national almanac, bringing in estimated revenues of $1,500.

29. McCulloch, J. R., and Martin, F., *A Dictionary Geographical, Statistical, and Historical, of the Various Countries, Places and Principal Natural Objects in the World*, vol. 3 (London, 1866), p. 432; Sky, T., *The National Road and the Difficult Path to Sustainable National Investment* (New York, 2011), p. 66.

30. Loomis, "Astronomical Observatories in the United States," p.26.

31. Dupree, A. H., *Science in the Federal Government: A History of Policies and Activities to 1940* (Cambridge, 1957), p. 62.

32. Loomis, "Astronomical Observatories in the United States," p. 26.

33. Portolano, M., "John Quincy Adams's Rhetorical Crusade for Astronomy," *ISIS* 91, no. 3 (2000), p. 49.

34. Quincy, J., *Memoir of the Life of John Quincy Adams* (Boston, 1860), p. 306.

35. Adams, J. Q., *An Oration Delivered Before the Cincinnati Astronomical Society on the Occasion of Laying the Corner Stone of an Astronomical Observatory on the 10th of November, 1843* (Cincinnati, OH, 1843), p. 68.

36. Zaban Jones, B., *Lighthouse of the Skies, The Smithsonian Astrophysical Observatory: Background and History, 1846–1955* (Washington, DC, 1965), p. 15.

37. Adams, *An Oration Delivered Before the Cincinnati Astronomical Society,* pp. 37–38.

38. Ibid., p. 17.

39. Quincy, *Memoir of the Life of John Quincy Adams,* p. 308.

40. Krisciunas, K., "A Short History of Pulkovo Observatory," *Vistas in Astronomy* 22 (1978), pp. 27–37.

41. Adams, *An Oration Delivered Before the Cincinnati Astronomical Society,* p. 60.

42. Sears, B., ed. "Report in the House of Representatives, March 1840, on the Smithsonian Bequest, from the Select Committee appointed on the subject," *Christian Review* 5 (Boston, 1840), p. 106. Estimate included "a salary of $3,600 for the astronomer, funds for the compensation of four assistants, at $1,500 each, and two laborers, each at $600; for the purchase and procurement of instruments, $30,000; of which $20,000 might be applied for an assortment of the best instruments to be procured, and $10,000 from a fund, from the interest of which other instruments may be from time to time procured, and for repairs: for the library, $30,000; being $10,000 for first supply, and $20,000 for a fund for an income of $1,200 a year: and finally $30,000 for a fund, from the income of which, $1,800 a year, shall be defrayed the expense of the yearly publication of the observations and of a nautical almanac."

43. Rhees, W. J., "Congressional Proceedings, Twenty-Sixth Congress, 1839–41," in *The Smithsonian Institution: Documents Relative to Its Origin and History* (Washington, DC, 1879), pp. 221–222.

44. Zaban Jones, *Lighthouse of the Skies,* p. 124.

45. Portolano, "John Quincy Adams' Rhetorical Crusade for Astronomy," p. 493.

46. Adams, J. Q., *The Great Design: Two Lectures on the Smithsonian Bequest by John Quincy Adams,* ed. W. Washburn (Washington, DC, 1965), p. 71.

47. Musto, "A Survey of the American Observatory Movement," p. 88.

48. Goodman, M., *The Sun and the Moon: The Remarkable True Account of Hoaxers, Showmen, Dueling Journalists and Lunar Man-Bats in Nineteenth-Century New York* (New York, 2008), pp. 133–134.

49. Locke, R. A., *The Moon Hoax; or, A Discovery That the Moon Has a Vast Population of Human Beings* (New York, 1859), pp. 36–37.

50. Ruskin, S. W., "A Newly-Discovered Letter of J. F. W. Herschel Concerning the 'Great Moon Hoax,'" *Journal for the History of Astronomy* 33, no. 110 (2002), p. 71.

51. Barnum, P. T., *The Humbugs of the World: An Account of Humbugs, Delusions, Impositions, Quackeries, Deceits and Deceivers Generally, in All Ages* (New York, 1866), p. 270.

52. Locke, *The Moon Hoax*, p. vi.

53. Crowe, M. J., *The Extraterrestrial Life Debate: Antiquity to 1915: A Source Book* (Notre Dame, 2008), p. 212.

54. Barnum, *The Humbugs of the World*, pp. 259–260.

55. Griggs, W. N., *The Celebrated "Moon Story," Its Origin and Incidents; With a Memoir of the Author and an Appendix* (New York, 1852), p. 16.

56. Goodman, *The Sun and the Moon*, p. 234.

57. Locke, *The Moon Hoax*, p. 15.

58. Ibid., p. 20.

59. Ibid., p. 15.

60. Adams, *An Oration Delivered Before the Cincinnati Astronomical Society*, p. 68.

61. Bruce, R., *The Launching of Modern American Science 1846–1876* (New York, 1987), p. 102.

62. Cobb, C., "Some Beginnings in Science," *Appleton's Popular Science Monthly*, October 1896, p. 767.

63. Williams, T., "Development of Astronomy in the Southern United States 1840–1914," *Journal for the History of Astronomy* 27, no. 1 (1996), p. 15.

64. Musto, "A Survey of the American Observatory Movement," p. 89.

65. Potts, D., *Wesleyan University, 1831–1910: Collegiate Enterprise in New England* (New Haven, CT, 1992), p. 78, http://www.wesleyan.edu/astro/vvo/history.html.

66. Prentice, G., *Wilbur Fisk* (Boston, 1890), p. 167.

67. Sewall, A., *Life of Prof. Albert Hopkins* (New York, 1870), p. 161.

68. Ibid., p. 162.

69. Cleminshaw, R. H., "Astronomy in the Early Days of the Western Reserve," *Popular Astronomy* 46 (1938), pp. 559–564.

70. Correspondence with Thomas Vince, Loomis Observatory historian and archivist; and Cleminshaw, "Astronomy in the Early Days of the Western Reserve," pp. 559–564.

71. Cutler, C., *A History of Western Reserve College, During its First Half Century, 1826–1876* (Cleveland, 1876), p. 46.

72. Cleminshaw, "Astronomy in the Early Days of the Western Reserve," p. 564.

73. Newton, H. A., *A Memoir of Elias Loomis* (Washington, DC, 1891), p. 762; This was the largest bequest that had ever been made to Yale at that time.

74. Musto, "A Survey of the American Observatory Movement," p. 89.

75. Loomis, "Astronomical Observatories in the United States," p. 29.

76. Edmonds, F. S., *History of the Central High School of Philadelphia* (Philadelphia, 1902), p. 91.

77. Ibid.

78. Ibid., p. 97.

79. Franklin, B., "Articles of Belief and Acts of Religion," in *The Works of Benjamin Franklin,* ed. J. Sparks, vol. 2 (Chicago, 1882), p. 1.

80. Newton, I., *The Mathematical Principles of Natural Philosophy,* vol. 2, ed. W. Davis and W. Emerson (London, 1803), pp. 310–311.

81. Rittenhouse, D., *An Oration Delivered February 24, 1775, Before the American Philosophical Society* (Philadelphia, 1775), p. 26.

82. Adams, *An Oration Delivered Before the Cincinnati Astronomical Society,* p. 17; Adams, *The Great Design,* p. 33.

83. Udias, A. *Searching the Heavens and the History of Jesuit Observatories* (New York, 2003).

84. Ibid., p. 147.

85. Ibid., p. 13.

86. Campbell, T. J., *The Jesuits, 1534–1921: A History of the Society of Jesus from Its Foundation to the Present Time* (New York, 1921), p. 852.

87. Curran, R. M., and O'Donovan, L. O., *The Bicentennial History of Georgetown University: From Academy to University 1789–1889* (Washington, DC, 1961), p. 143; Udias, *Searching the Heavens and the History of Jesuit Observatories,* p. 104.

88. Curran and O'Donovan, *The Bicentennial History of Georgetown University,* p. 141.

89. Ibid.

90. Longstreth, R., "Biographical Memoir of Miers Fisher Longstreth 1819–1891," *National Academy of Sciences Biographical Memoirs,* vol. 8 (1915), pp. 137–140.

91. Dupree, *Science in the Federal Government,* p. 62; Dick, S. J., *Sky and Ocean Joined: The U.S. Naval Observatory 1830–2000* (Cambridge, 2003).

92. Loomis, "Astronomical Observatories in the United States," p. 35.

93. Dick, *Sky and Ocean Joined,* p. 28.

94. Ibid., p. 27.

95. Dick, S. J., "John Quincy Adams, the Smithsonian Bequest and the Founding of the U.S. Naval Observatory," *Journal for the History of Astronomy* 22, no. 1 (1991), p. 41.

96. Loomis, "Astronomical Observatories in the United States," p. 32. The cost of the Merz and Mahler 9.5 aperture equatorial was $6,000; the Merz and Mahler/Ertel and Son 5.5-inch transit instrument, $1,480; the Simms 5-foot mural circle, $3,550; the Pistor and Martins 5-inch transit, $1,750; and the 4-inch Merz and Mahler comet seeker, $280. Dick, *Sky and Ocean Joined,* p. 53.

97. Maury, M., *Astronomical Observations Made During the Year 1845 at the U.S. Naval Observatory, Washington* (Washington, DC, 1846).

98. Dick, *Sky and Ocean Joined,* p. 57.

99. Ibid., p. 207.

100. Loomis, "Astronomical Observatories in the United States," p. 35.

2. Public Spirit and Patronage

1. Goode, G. B., "The Origin of the National Scientific and Educational Institutions of the United States," *Annual Report of the American Historical Association for the Year 1889* (Washington, DC, 1890), pp. 53–163.

2. Miller, H., *Dollars for Research: Science and Its Patrons in Nineteenth-Century America* (Seattle, 1970), p. 29.

3. U.S. Census of 1840 and 1850, as seen at http://www.census.gov/population/www/documentation/twps0027/tab07.txt and https://www.census.gov/population/www/documentation/twps0027/tab08.txt.

4. Miller, *Dollars for Research*, p. 30.

5. Ibid.

6. Cleminshaw, C. H., "The Founding of the Cincinnati Observatory," *Astronomical Society of the Pacific, Leaflets* 5 (1946), p. 68.

7. Black, R., "The Cincinnati Telescope," *Popular Astronomy*, Vol. 52 (1944), p. 77.

8. "Constitution of the Cincinnati Astronomical Society," in Adams, J. Q., *An Oration Delivered Before the Cincinnati Astronomical Society on the Occasion of Laying the Corner Stone of an Astronomical Observatory on the 10th of November, 1843* (Cincinnati, 1843), p. 70.

9. Cleminshaw, "The Founding of the Cincinnati Observatory," p. 68.

10. Miller, *Dollars for Research*, p. 30.

11. Ibid., p. 33.

12. Black, "The Cincinnati Telescope," p. 78.

13. Adams, *An Oration Delivered Before the Cincinnati Astronomical Society*, p. 63.

14. Miller, *Dollars for Research*, p. 32.

15. Bailey, S., *The History and Work of Harvard Observatory: 1839 to 1927* (New York, 1931), p. 15.

16. Ibid., p. 24.

17. Rothenberg, M., "Patronage of Harvard College Observatory 1839–1851," *Journal for the History of Astronomy* 21, no. 1 (1990), p. 37.

18. Rothenberg, "Patronage of Harvard College Observatory," p. 39.

19. Bailey, S., *The History and Work of Harvard Observatory*, p. 26. The observatory's fifteen-inch equatorial telescope alone cost $19,842.

20. Miller, *Dollars for Research*, p. 38.

21. Bailey, *The History and Work of Harvard Observatory*, p. 22.

22. Wise, G., *Civic Astronomy: Albany's Dudley Observatory, 1852–2002* (New York, 2004), p. 13.

23. See James, M. A., *Elites in Conflict: The Antebellum Clash over the Dudley Observatory* (Camden, NJ, 1987).

24. Miller, *Dollars for Research*, pp. 41–43.

25. Wise, *Civic Astronomy: Albany's Dudley Observatory*, pp. 9–10.

26. The Trustees of the Dudley Observatory, *Dudley Observatory and the Scientific Council: Statement of the Trustees* (Albany, 1858), p. 74.

27. Gould, B., *Reply to the "Statement of the Trustees" of the Dudley Observatory* (Albany, 1859), p. 55.

28. Ibid., p. 98.

29. Loomis, E., "Astronomical Observatories in the United States," *Harper's New Monthly Magazine* 13 (June 1856), p. 47. A significant share of the cost was contributed personally by the college president, Rev. William J. Walker.

30. Williams, T., "The Development of Astronomy in the Southern United States, 1840–1914," *Journal for the History of Astronomy* 27, no. 1 (1996), p. 21.

31. Whitesell, P., *A Creation of His Own: Tappan's Detroit* (Ann Arbor, MI, 1998), pp. 29–32.

32. Dwight, B. W., *The History of the Descendants of Elder John Strong of Northampton, Mass.* (Albany, 1871), p. 377; Loomis estimated the building to have cost $5,000 and the refractor, $10,000.

33. Anonymous, "Frederick Augustus Porter Barnard," *Proceedings of the American Academy of Arts and Sciences,* vol. 24 (1889), p. 441.

34. Byrd, G., and R. Mellown, "An Antebellum Observatory in Alabama," *Sky and Telescope,* February 1983, p. 113; American Institute of the City York, *Transactions of the American Institute of the City of New York for the Year 1858* (Albany, 1859), p. 417.

35. Williams, "The Development of Astronomy in the Southern United States," p. 17.

36. Sansing, D., *The University of Mississippi: Sesquicentennial History* (Oxford, 1999), pp. 91–92.

37. Ibid., p. 91.

38. Beardsley, W., "The Allegheny Observatory During the Era of the Telescope Association, 1859–1867," *Western Pennsylvania Historical Magazine* 64 (1981), p. 216.

39. Ibid., p. 222.

40. Brashear, J., "An Address Delivered at the Laying of the Cornerstone of the New Observatory, October 20, 1900," *Miscellaneous Scientific Papers of the Allegheny Observatory of the University of Pittsburgh,* vol. 2 (Lancaster, 1910), p. 34.

41. Ibid., p. 8; Allegheny Centennial Committee, *Story of Old Allegheny City* (Pittsburgh, 1941), p. 164.

42. Colbert, E., "The Early Years of the Dearborn Observatory," *Popular Astronomy* 24 (1916), p. 476.

43. Winborne, B. B., *The Colonial and State Political History of Hertford County, N.C.* (Raleigh, 1906), p. 182.

44. Colbert, "The Early Years of the Dearborn Observatory," p. 477.

45. Williams, "Development of Astronomy in the Southern United States," pp. 23–25.

46. Colbert, "The Early Years of the Dearborn Observatory," p. 477.

47. Goodspeed, T. W., *A History of the University of Chicago* (Chicago, 1916), p. 15.

48. For comparison, in 2015, *Forbes* magazine listed 1,826 billionaires around the world, 536 of whom were from the United States; Ratner, S., *New Light on the History of Great American Fortunes: American Millionaires of 1892 and 1902* (New York, 1953). Also see Piketty, T., and Saez, E., "Income Inequality in the United States, 1913–1998," *Quarterly Journal of Economics* 108, no. 1 (2003), pp. 1–39.

49. Loomis, "Astronomical Observatories in the United States," p. 46; and Dartmouth College, "History of Dartmouth College (George C. Shattuck) Observatory" (1995), http://ead.dartmouth.edu/html/da9.html. Loomis says the building cost $4,500; the equatorial and sidereal clock, $2,300; comet seeker, $18; and meridian circle, £275.

50. The Regents of the University, *University of the State of New York: Ninety-Second Annual Report of the Regents of the University* (Albany, 1879), pp. 195–196.

51. Williams, "The Development of Astronomy in the Southern United States," p. 26.

52. Ibid., p. 28.

53. Ibid., p. 31.

54. University of Virginia, Board of Visitors, *Board of Visitors Minutes,* University of Virginia Library Digital Collections (Charlottesville, 2006), p. 93.

55. Williams, J. R., *The Handbook of Princeton* (New York, 1905), p. 49.

56. Ibid.

57. Kelly, B. M., *Yale: A History* (New Haven, CT, 1999), p. 206.

58. Pritchett, C., "Morrison Observatory, Glasgow, Mo.," *Annual Record of Science and Industry for 1877,* ed. S. F. Baird (New York, 1878), p. 44.

59. Brown, D., *The Story of Morrison Observatory: 100 Years* (Fayette, 1975).

60. Morrison-Fuller, B., *Plantation Life in Missouri* (Glasgow, 1937), p. 31.

61. Lick, R., *The Generous Miser: The Story of James Lick of California* (Los Angeles, 1967), p. 62.

62. Wright, H., *James Lick's Monument: The Saga of Captain Richard Floyd and the Building of the Lick Observatory* (Cambridge, 1987), p. 33.

63. Gingerich, O. J., ed., *Astrophysics and Twentieth-Century Astronomy to 1950, Part A: The General History of Astronomy* (New York, 1984), p. 127; Miller, *Dollars for Research,* p. 103. Davidson had initially suggested that $1.2 million would be required, and Lick had initially only been willing to spend $500,000. Lick settled on $700,000.

64. Ashbrook, J., *Astronomical Scrapbook: Skywatchers, Pioneers, and Seekers in Astronomy* (Cambridge, 1984), pp. 74–75.

65. Ibid. When the naked-eye comet of August 1881 brought three thousand letters claiming the prize, Warner changed the terms of the prize to the best essay on comets.

66. Manning, M., *Man, Mountain and Monument: An Historical Account of Professor Thaddeus S. C. Lowe and the Mount Lowe Railway* (Altadena, 2001). Also, see

Winter, F., "The 'Trip to the Moon' and Other Early Spaceflight Simulation Shows, ca. 1901–1915: Part I," in *History of Rocketry and Astronautics* 23, no. 1 (2001), pp. 133–162.

67. Bless, R., "Washburn Observatory," paper, 1978, available at http://www .astro.wisc.edu/~varda/Long_Wash_Obs_Text.html.

68. Ibid.

69. Upton, W., "The Ladd Observatory," *Sidereal Messenger* 10 (1891).

70. Brown, C., "From Boom to Bust: Humphrey Barker Chamberlin, 1880–1894," paper, 1980, p. 8, available at www.denverastrosociety.org/dfiles/chambio.pdf.

71. Leonard, D. L., *The History of Carleton College: Its Origin and Growth, Environment and Builders* (Chicago, 1904), pp. 233–234.

72. Menke, D., "Dinsmore Alter and the Griffith Observatory," *Planetarian* 16, no. 4 (1987).

73. Loomis, "Astronomical Observatories in the United States," p. 42. The 6.3-inch Merz equatorial cost $1,833, and a Young meridian, $800.

74. For the tradition of the "Grand Amateur" astronomer in Great Britain, see Chapman, A., *The Victorian Amateur Astronomer: Independent Astronomical Research in Britain 1820–1920* (London, 1999).

75. National Academy of Sciences, "Memoir of Lewis Morris Rutherfurd," *National Academy of Sciences: Biographical Memoires,* vol. 3 (Washington, DC, 1895), pp. 417–441; Warner, D., "Lewis M. Rutherfurd: Pioneer Astronomical Photographer and Spectroscopist," *Technology and Culture* 12 (1971), pp. 190–216.

76. Abrams, P., "Henry Fitz, American Telescope Maker," *Journal of the Antique Telescope Society* 6 (1994).

77. Barker, G., *Memoir of Henry Draper: 1837–1882* (New York, 1888).

78. Loomis, "Astronomical Observatories in the United States," p. 46.

79. A comprehensive description of American telescope builders and their development falls outside of the scope of this study. American astronomical observatories provided a source of domestic demand that stimulated technical progress among the builders, whose activities then served as enablers of further observatory projects as well as popular astronomy more generally. By the end of the century, American telescope makers, notably the firm Alvan Clark & Sons, were among the best in the world, particularly in the domain of larger refractors. While astronomers and their patrons were often the motivating forces behind American observatories, it was the telescope builders who provided the technical expertise to construct them. Intrinsic motivations were core to telescope making as well: Ambrose Swasey—one half of the precision machining partnership of Warner & Swasey, whose telescope-making careers included building the mechanisms and mountings for the telescopes of the Lick, Yerkes, McDonald, and Dominion Astrophysical Observatories—said of their business, "we get our money out of machinery and our glory out of telescopes." For more on nineteenth-century American telescope makers, see Abrahams, P., "Henry Fitz, American Telescope Maker," *Journal of the Antique Telescope Society* 6 (Summer 1994), pp. 6–10; Bell, T., "In the Shadow of Giants: Forgotten Nineteenth-

Century Telescope Makers and Their Crucial Role in Popular Astronomy," *Griffith Observer* 50, no. 9 (September 1986), pp. 3–14; Bell, T., "Money and Glory," *The Bent of Tau Beta Pi* (Winter 2006), pp. 13–20; Bagdasarian, N., "Amasa Holcomb: Yankee Telescope Maker," *Sky & Telescope,* June 1986, pp. 620–622; Brashear, J., *John A. Brashear: The Autobiography of a Man Who Loved the Stars,* ed. W. L. Scaife (New York, 1924); Preston, F. W., and McGrath Jr., W. J., *Holcomb, Fitz, and Peate: Three 19th Century American Telescope Makers* (Washington, DC, 1962); Warner, D. J., and Ariail, R. B., *Alvan Clark & Sons: Artists in Optics* (Richmond, 1995).

80. Putnam, W. L., *The Explorers of Mars Hill: A Centennial History of Lowell Observatory 1894–1994* (Flagstaff, 1994), p. 12.

81. See Lowell, P., *Mars and Its Canals* (New York, 1906) and *Mars as the Abode of Life* (New York, 1908).

82. Strauss, D., *Percival Lowell: The Culture and Science of a Boston Brahmin* (Cambridge, 2001), p. 54.

83. Ibid.

84. Ibid., p. 55.

85. Joy, J., ed., "The Giver of the Perkins Telescope," *Christian Advocate* 97, no. 26 (1922), pp. 806–807.

86. Crump, C., "The Perkins Observatory of the Ohio Wesleyan University," *Popular Astronomy* 37, no. 10 (1929), p. 552. The telescope pier and instruments cost $93,580; the sixty-one-inch mirror, $33,316; the mounting, $9,000; the building, $152,000.

87. Evans, D. S., and Mulholland, J. D., *Big and Bright: A History of the McDonald Observatory* (Austin, 1986), pp. 9–11.

88. Ibid., p. 20.

89. Florence, R., *The Perfect Machine: Building the Palomar Telescope* (New York, 1994), p. 26.

90. "Recollections of Childhood," August 5, 1935, box 92, Biographical Notes, Papers of George Ellery Hale, Archives, California Institute of Technology.

91. Ibid.

92. Osterbrock, D., *Pauper & Prince: Ritchey, Hale, & Big American Telescopes* (Tucson, 1993), p. 22.

93. Ibid., p. 25.

94. Ibid., p. 68.

95. Hendrick, B., *The Age of Big Business: A Chronicle of the Captains of Industry* (New Haven, CT, 1921), p. 126.

96. Miller, H., "Astronomical Entrepreneurship in the Gilded Age," *Astronomical Society of the Pacific Leaflet,* no. 479 (May 1969), p. 3.

97. Wright, H., Warnow, J., and Weiner, C., *The Legacy of George Ellery Hale* (Cambridge, 1972), p. 19.

98. Wright, H., *Explorer of the Universe* (New York, 1966), pp. 98–99.

99. Lerner, R., *Astronomy Through the Telescope* (Toronto, 1982), p. 104.

100. Miller, *Dollars for Research,* p. 110.

101. "Yerkes Will Founds Gallery and Hospital," *New York Times,* January 3, 1906, p. 1; Franch, J., *Robber Baron: The Life of Charles Tyson Yerkes* (Chicago, 2006), p. 323.

102. Miller, *Dollars for Research,* p. 109.

103. Osterbrock, *Pauper & Prince,* p. 34.

104. Carnegie, A., "Trust Deed by Andrew Carnegie," *The Carnegie Institution of Washington: Founded by Andrew Carnegie* (Washington, DC, 1902), p. 8.

105. Osterbrock, *Pauper & Prince,* p. 71.

106. Florence, *The Perfect Machine,* p. 37.

107. Carnegie, A., *The Gospel of Wealth and Other Timely Essays* (New York, 1901), p. 25.

108. Plotkin, H., "Edward C. Pickering and the Endowment of Scientific Research in America, 1877–1918," *ISIS* 69 (1978), p. 51.

109. Wright, *Explorer of the Universe,* p. 253.

110. Hale, G., "The Possibilities of Large Telescopes," *Harper's Monthly* 79 (1928), p. 640.

111. Carnegie Institution, *Carnegie Institution of Washington: Year Book No. 20, 1921* (Washington, DC, 1922), p. 40. The year 1912, the middle year between the 1906 commencement of work on the mirror blank and before the 1917 completion of the hundred-inch reflector, is used as the base year for calculation. For comparison, the next highest capital expenditure for the Carnegie Institution was for its "Department of Terrestrial Magnetism" at a cost of $443,000.

112. Letter from Carnegie to Hale, November 19, 1906, and Letter from Hale to Carnegie, December 11, 1906, box 10, "Andrew Carnegie," Papers of George Ellery Hale, Archives, California Institute of Technology.

113. Letter to Carnegie from Hale, May 5, 1910, in ibid.

114. Letter to Carnegie from Assistant, July 2, 1910, in ibid.

115. Letter to Hale from Carnegie, February 14, 1913, in ibid.

116. Letter to Mrs. Carnegie from Hale, September 21, 1921, in ibid.

117. Undated and unidentified newspaper clipping, in ibid.

118. Carnegie, A., *Autobiography of Andrew Carnegie* (London, 1920), pp. 261–262.

119. Van Helden, A., "Building Large Telescopes," in *Astrophysics and Twentieth Century Astronomy to 1950, Part A: The General History of Astronomy,* ed. O. J. Gingerich (New York, 1984), p. 138.

120. Geiger, R., *To Advance Knowledge: The Growth of American Research Universities, 1900–1940* (New Jersey, 2004), p. 162.

121. Hale, "The Possibilities of Large Telescopes," p. 639.

122. Ibid.

123. Wright, *Explorer of the Universe,* p. 390.

124. Florence, *The Perfect Machine,* pp. 91, 388. Given that it would be over twenty years until the project was completed, it is difficult, without a more complete

expenditure time series, to state a precise equivalent value today. The 2015 GDP-ratio equivalent value of using the 1949 end date would be $433 million, and using 1934, the year of site selection and mirror casting, yields a value of $2.07 billion, due to the economic collapse during the Depression.

125. Wright, *Explorer of the Universe*, p. 390.

126. Letter from Rose to Hale, November 7, 1928, and Letter from Rose to Hale, November 14, 1929, box 35, "Wickliffe Rose 1924–1932," Papers of George Ellery Hale, Archives, California Institute of Technology.

127. Fosdick, R., *John D. Rockefeller, Jr.: A Portrait* (New York, 1956), p. 373.

128. Florence, *The Perfect Machine*, p. 73.

129. Osterbrock, *Pauper & Prince*, p. 227.

130. Letter from Hale to Robinson, April 8, 1915, box 35, "Henry M. Robinson 1915–1937," Papers of George Ellery Hale, Archives, California Institute of Technology.

131. Letter from Robinson to Hale, December 29, 1923, and Letter from Robinson to Hale, July 26, 1929, in ibid.

132. Florence, *The Perfect Machine*.

133. Edmondson, F., *Aura and Its US National Observatories* (Cambridge, 1997).

134. Browne, M., "Giant Telescope Prepares to Take Its First Look Skyward," *New York Times*, January 9, 1990, p. C1.

135. Nielsen, W., *Golden Donors* (New York, 1985), p. 234.

136. Lankford, J., *American Astronomy* (Chicago, 1997).

137. Plotkin, H., "Edward C. Pickering and the Endowment of Scientific Research in America, 1877–1918," p. 46.

138. Stroobant, P., Delvosal, J., Philippot, H., Delport, E., and Meelin, E., *Les Observatories Astronomiques et les Astronomes* (Brussels, 1907), pp. 3–4.

139. McKenney, A., "What Women Have Done for Astronomy," *Popular Astronomy* 12 (1904), pp. 171–182.

140. Hale, G., and Keeler, J., eds., "Catherine Wolfe Bruce," *Astrophysical Journal* 11 (1900), p. 168.

141. Ibid., p. 116.

3. Spaceflight, Millionaires, and National Defense

1. Senate Committee on Aeronautical and Space Sciences, *Congressional Recognition of Goddard Rocket and Space Museum Roswell, New Mexico* (Washington, DC, 1970), p. 24.

2. This chapter builds on the extensive scholarship of many previous authors who have investigated the life and works of Robert Goddard. Particularly, Clary, D., *Rocket Man: Robert H. Goddard and the Birth of the Space Age* (New York, 2003); Lehman, M., *This High Man: The Life of Robert H. Goddard* (New York, 1963); Pendray, E. G., "Pioneer Rocket Development in the United States," *Technology and Culture* 4, no. 4 (1963); Dewey, A. P., *Robert Goddard: Space Pioneer* (Boston, 1962); Verral, C. S., *Robert Goddard: Father of the Space Age* (Englewood Cliffs, NJ, 1963);

Rhodes, R., "The Ordeal of Robert Hutchings Goddard: 'God Pity a One-Dream Man,'" *American Heritage* 31, no. 4 (1980); Hacker, B. C., "Robert H. Goddard and the Origins of Space Flight," in *Technology in America: A History of Individuals and Ideas,* ed. C. W. Pursell Jr. (Cambridge, MA, 1981); Hunley, J. D., "The Enigma of Robert H. Goddard," *Technology and Culture* 36, no. 2 (1995), pp. 327–350; Durant, F. C., "Robert H. Goddard and the Smithsonian Institution," in *First Steps Toward Space,* ed. F. C. Durant III and G. S. James (Washington, DC, 1974), pp. 57–69; Durant, F. C., "Robert H. Goddard: Accomplishments of the Roswell Years, 1930–1941," in *History of Rocketry and Astronautics,* ed. K. R. Lattu, vol. 8 (San Diego, 1989), pp. 317–341; Durant, F. C., and Winter, F., "Goddard and Lindbergh: The Role of Charles A. Lindbergh in the Rocketry Career of Robert H. Goddard," in *History of Rocketry and Astronautics,* ed. M. L. Ciancone, vol. 33 (San Diego, 2010), pp. 31–59; Siddiqi, A. A., "Deep Impact: Robert Goddard and the Soviet 'Space Fad' of the 1920s," *History and Technology* 20, no. 2 (2004).

3. Goddard, R., *The Papers of Robert H. Goddard,* vol. 1 (New York, 1970), pp. 7–8.

4. Serviss, G. P., *Edison's Conquest of Mars* (Boston, 1898).

5. Goddard, *The Papers of Robert H. Goddard,* vol. 1, p. 7.

6. Serviss, *Edison's Conquest of Mars,* cap. 3.

7. Kuznets, S., *Capital in the American Economy* (Princeton, 1961), p. 557.

8. Panama Canal Authority, "A History of the Panama Canal: French and American Construction Efforts," http://www.pancanal.com/eng/history/history/end.html.

9. Lehman, *This High Man,* p. 17.

10. Ibid., p. 29.

11. Goddard, *The Papers of Robert H. Goddard,* vol. 1, p. 117.

12. Ibid., vol. 3, pp. 1611–1612.

13. Ibid., vol. 1, p. 420.

14. Wilford, J., "The Moon May Have Water, and Many New Possibilities," *New York Times,* December 4, 1996; Leary, W., "NASA Plans Permanent Moon Base," *New York Times,* December 5, 2006.

15. Goddard's multiple-charge solid-fuel-rocket design was unique: it was a rocket that was to be fed dozens of cartridges of solid-fuel propellant that would be expended and then discarded in sequence by an automated loading device in flight. The practical difficulties of making this device work ultimately led Goddard to abandon the years of research and development that he had expended on it and to focus instead on liquid-fuel propulsion.

16. Goddard, *The Papers of Robert H. Goddard,* vol. 1, pp. 161, 166.

17. Even his long-standing assistant and instrument maker, Nils Riffolt, was unaware that his master's thesis, "A Study in the Absorption of Radiant Energy," helped provide validation for Goddard's plans for using solar energy for in-space propulsion. Ibid., pp. 38–39.

18. Ibid., pp. 126–127.

19. Ibid., p. 164.

20. Ibid., p. 171.

21. Ibid.

22. Ibid., p. 175.

23. Ibid., p. 174.

24. See Smithson, J., "Will of James Smithson," in *Smithsonian Miscellaneous Collections*, vol. 17 (Washington, DC, 1880), pp. 1–2.

25. Goddard, *The Papers of Robert Goddard*, vol. 1, pp. 175, 181.

26. Ibid., p. 190. Thomas Hodgkins, a New York candy-store magnate, had provided the Smithsonian with $200,000 in 1891 with specifications that $100,000 of it be dedicated to research on the nature of the properties of air. The Smithsonian's generous interpretation of Hodgkins's intent to include the development of high-altitude rocketry is an example of how loosely defined funding criteria for philanthropic institutions can enable the funding of innovative projects.

27. Goddard, *The Papers of Robert Goddard*, vol. 1, p. 194. It would also be these attributes that would be touted by the promoters of the V-2 and that were realized by it in the Second World War.

28. Ibid., p. 194.

29. Ibid.

30. Ibid., p. 195.

31. Ibid., p. 196.

32. Ibid., pp. 199–201.

33. See U.S. House of Representatives, Committee on Military Affairs, Subcommittee on Aviation, *United Air Service: Hearing Before a Subcommittee of the Committee on Military Affairs*, 66th Cong., 2nd sess. (1921), pp. 191–206; Van der Linden, F. R., *Airlines and Air Mail: The Post Office and the Birth of the Commercial Aviation Industry* (Lexington, 2002), p. 78.

34. Goddard, R., "Diary, August 28, 1917," *Goddard Digital Collection*, Robert Hutching Goddard Library, Clark University, Worcester, MA; Dunwoody, H., *Notes, Problems and Laboratory Exercises in Mechanics, Sound, Light, Thermo-Mechanics and Hydraulics Prepared for Use in Connection with the Course in Natural and Experimental Philosophy at the United States Military Academy* (New York, 1917).

35. Goddard, "Diary, August 28, 1917."

36. Ibid.

37. Goddard, "Diary, October 24, 1917," and "Diary, November 11, 1917."

38. Goddard, "Diary, November 12, 1917," and "Diary, November 13, 1917."

39. Goddard, "Diary, December 5, 1917"; House of Representatives, Committee on Military Affairs, Subcommittee on Aviation, *United Air Service*, p. 191.

40. Goddard, "Diary, January 17–22, 1918."

41. Walcott had sent Stratton, the director of the National Bureau of Standards, Goddard's first proposal for evaluation. Edgar Buckingham, a physicist at the bureau, who acted with Abbot as the supervisory committee for Goddard's work during the war, had performed the actual evaluation.

42. For more on this and Squier, see Gross, C., "George Owen Squier and the Origins of American Military Aviation," *Journal of Military History* 54, no. 3 (1990), pp. 281–306.

43. Ibid., p. 291.

44. The report of Abbot and Buckingham emphasized the long-range aspects, noting that "enormous distances, far outranging rifled cannon, can be reached" and suggesting that a 120-mile range—roughly half of the V-2's operational range—might be possible; Goddard, *The Papers of Robert Goddard,* vol. 1, p. 211.

45. Neufeld, M. J., *The Rocket and the Reich: Peenemünde and the Coming of the Ballistic Missile* (New York, 1995), pp. 41–53; Bainbridge, W. S., *The Spaceflight Revolution: A Sociological Study* (New York, 1976), pp. 45–85.

46. Gross, C., "George Owen Squier," p. 300.

47. Goddard, *The Papers of Robert Goddard,* vol. 1, pp. 226, 249, 277.

48. Goddard, "Diary, August 28, 1917."

49. Goddard, "Diary, February 13, 1917."

50. Goddard, "Diary, November 21, 1917."

51. Washburn, C., *Industrial Worcester* (Worcester, 1917), p. 283.

52. Goddard, *The Papers of Robert Goddard,* vol. 1, p. 205; Goddard, "Diary, December 15–January 15, 1917."

53. Goddard, *The Papers of Robert Goddard,* vol. 1, pp. 230–232.

54. Ibid., p. 232n.; Goddard, "Diary, July 25–26, 1919," and "Diary, January 16, 1920."

55. Goddard, *The Papers of Robert Goddard,* vol. 1, p. 232.

56. Ibid., p. 294.

57. Ibid., pp. 255, 265–276.

58. Ibid., p. 254.

59. Ibid., pp. 300–301.

60. Ibid., pp. 301–302.

61. Lehman, *This High Man,* pp. 121–123.

62. Unmentioned in Lehman; one sentence in Clary, *Rocket Man,* p. 84.

63. Goddard, *The Papers of Robert Goddard,* vol. 1, pp. 244, 253, 278, 283.

64. Ibid., p. 441.

65. Ibid., pp. 443–444.

66. Ibid., p. 446.

67. Ibid., p. 447.

68. Goddard, "Diary, August 7, 1920."

69. Goddard, "Diary, November 16, 1920,"; Lehman, *This High Man,* pp. 124–125; Goddard, "Diary, April 9–19, 1923."

70. Goddard, *The Papers of Robert Goddard,* vol. 1, p. 488.

71. Goddard, "Diary, February 18, 1923," and "Diary, May 8–11, 1923"; Lehman, *This High Man,* p. 122; Goddard, *The Papers of Robert Goddard,* vol. 1, pp. 488, 496.

72. For contemporary accounts of the effects of gas attacks, see; Fries, A., and West, C., *Chemical Warfare* (New York, 1921), pp. 13–14; New York Times, *Current History: The European War,* vol. 12 (New York, 1917), p. 125.

73. Goddard, *The Papers of Robert Goddard,* vol. 1, pp. 244, 432, 465.

74. Oberth also reports that his position with a field ambulance unit in the Austro-Hungarian Army also led to other early space experiments. Apparently, his position allowed him to study pharmacy, which led him to experiment with the psychotropic substance scopolamine, an extract from the plant henbane, and to immerse himself in a tub of water in order to simulate weightlessness. Hartl, H., *Herman Oberth—Vorkaempfer Der Weltraumfahr* (Hanover, 1958), p. 77; Walters, H., *Hermann Oberth: Father of Space Travel* (New York, 1962), p. 33.

75. Goddard, *The Papers of Robert Goddard,* vol. 1, p. 316.

76. Ibid., p. 308.

77. Ibid., p. 320.

78. Ibid., p. 322.

79. Ibid., p. 329.

80. Goddard, "Diary, July 25–26, 1919," "Diary, August 20–21, 1919," and "Diary, January 16, 1920."

81. David Clary's *Rocket Man* has provided the most thorough account of Goddard's need for patrons and his patronage relationships.

82. Goddard, *The Papers of Robert Goddard,* vol. 1, p. 393.

83. Ibid.

84. "S'pose Prof. Goddard's Rocket Does Hit Moon," *New Haven Register,* January 13, 1920; "Great Scheme, Beats Soviet Ark and Note Writing for Getting Rid of Things," *Philadelphia Inquirer,* January 13, 1920; "If It Hits the Moon We'll Get Moonshine Back," *New York Evening Journal,* January 13, 1920—all in *Goddard Digital Collection,* Robert Hutching Goddard Library, Clark University, Worcester, MA.

85. "Atmosphere Edge Conditions to Be Studied by Rocket," *Newark News,* January 12, 1920, *Goddard Digital Collection.*

86. "Sims Thinks Rocket Can Be Shot to Moon," *Rochester Post Express,* January 16, 1920, *Goddard Digital Collection.*

87. "Off the Earth," *St. Louis Star,* January 16, 1920, *Goddard Digital Collection.*

88. "Not a Moon Rocket," *Pittsburgh Dispatch,* January 17, 1920, *Goddard Digital Collection.*

89. Collins, T., *The Legendary Model T Ford: The Ultimate History of America's First Great Automobile* (Iola, WI, 2007), p. 88.

90. Goddard, *The Papers of Robert Goddard,* vol. 1, p. 410.

91. Ibid., p. 409.

92. "Photographing the Moon," *Baltimore News,* January 21, 1920, *Goddard Digital Collection.*

93. "Wants Pay Passenger for Voyage to Moon," *Newark Star Eagle,* January 31, 1921, *Goddard Digital Collection.*

94. "Chance to Invest," *Bridgeport Telegram,* January 28, 1920, *Goddard Digital Collection.*

95. Ibid.

96. Astor, J. J., *A Journey in Other Worlds* (New York, 1894).

97. Shaw, A., ed., "A Journey in Other Worlds," *Review of Reviews* 9 (January–June, 1894), p. 753. For a more critical and amusing review from Boston, see Anonymous, "A Journey in Other Worlds," *Literary World* 25, no. 10 (1894), p. 149.

98. "Proposes to Leap to Mars," *Boston Herald,* February 5, 1920, *Goddard Digital Collection.*

99. "Four Seek Trip to Mars," *New York Times,* February 9, 1920, *Goddard Digital Collection.*

100. Lehman, *This High Man,* pp. 109–112.

101. "That Flight to Planet Mars," *Spokesman Review,* February 25, 1920, *Goddard Digital Collection.*

102. Clary, *Rocket Man,* pp. 92–97.

103. Goddard, *The Papers of Robert Goddard,* vol. 1, p. 71.

104. This was the Dearborn Observatory. For more on its history, see chapter 2.

105. "Seek Moon Trip Fund in Chicago," *Chicago Herald,* April 3, 1920, and "Chicago Is Ready to Push Rocket to Moon," *Boston Globe,* April 11, 1920, *Goddard Digital Collection.*

106. "Windy City Capitalists May Finance Work on Rocket to Reach Moon," *Worcester Sunday Telegram,* April 11, 1920, and "Chicago Is Ready to Push Rocket to Moon," *Boston Globe,* April 11, 1920, *Goddard Digital Collection.*

107. "If you can't be present at dinner," Chicago Chapter American Association of Engineers Ladies' Night Poster, April 2, 1920, *Goddard Digital Collection.*

108. Goddard, "Diary, April 7, 1920," and "Diary, April 14, 1920," *Goddard Digital Collection.*

109. The effects of this announcement in Russia are discussed in Siddiqi, A. A., "Deep Impact"; Goddard, "Diary, April 15, 1920," and "Diary, April 19, 1920," *Goddard Digital Collection.*

110. "Goddard Rocket to Shoot in July," *Boston Herald,* April 28, 1920, and "Rocket Trip to Moon Postponed to August," *Washington Times,* July 15, 1920, *Goddard Digital Collection.*

111. "To Bombard the Moon," *Boston Post,* January 15, 1920, and "War Rocket 200-Mile Range," *Detroit Free Press,* September 5, 1920, *Goddard Digital Collection.*

112. "Only 'Angle' Needed for Trip to Moon," *Boston Sunday Advertiser,* September 19, 1920, *Goddard Digital Collection.*

113. Winter, F., *Prelude to the Space Age: The Rocket Societies 1924–1940* (Washington, DC, 1983), p. 73.

114. "Goddard's Rocket in War," *Buffalo Inquirer,* January 16, 1920, *Goddard Digital Collection.*

115. "Needs Money to Make Journey to the Moon," *Washington Times,* September 20, 1920, *Goddard Digital Collection.*

116. "Only 'Angel' Needed for Trip to Moon," *Boston Sunday Advertiser,* September 19, 1920, *Goddard Digital Collection.*

117. "That Earth-to-the-Moon Scheme," *Troy Times,* October 5, 1920, *Goddard Digital Collection.*

118. "Goddard's Rocket Is Not Practical," *Evening Gazette,* December 28, 1920, *Goddard Digital Collection.*

119. Goddard, *The Papers of Robert Goddard,* vol. 1, pp. 471–472.

120. Ibid., p. 470.

121. "Clark to Finance Goddard's Rocket," *Evening Gazette,* undated, 1921, *Goddard Digital Collection.*

122. Goddard, *The Papers of Robert Goddard,* vol. 1, p. 477.

123. Clarke, A. C., *The Exploration of Space* (London, 1951), p. 27.

124. From the third edition of the work. Oberth, H., *Wege zu Raumschiffahrt,* trans. Agence Tunisienne de Public-Relations (Washington, DC, 1970), p. 1.

125. For more on the early culture of spaceflight in Germany, see Neufeld, M., "Weimar Culture and Futuristic Technology," *Technology and Culture* 31, no. 4 (1990); Essers, I., *Max Valier: A Pioneer of Space Travel* (Washington, DC, 1976); Winter, *Prelude to the Space Age.*

126. Goddard, *The Papers of Robert Goddard,* vol. 1, p. 498.

127. Ibid., p. 519.

128. Goddard had given a talk at the annual conference of the American Association for the Advancement of Science that year in Cincinnati and had previously been pursuing a $3,000 grant from them. Ibid., pp. 521, 524, 528.

129. Bush, V., *Frederick Gardner Cottrell: 1877–1948* (Washington, DC, 1952).

130. Ibid., pp. 1–2.

131. Goddard, *The Papers of Robert Goddard,* vol. 1, p. 524.

132. Ibid., p. 543.

133. Ibid., vol. 2, p. 580.

134. Ibid.

135. Ibid., p. 591.

136. Ibid., p. 596.

137. Ibid., pp. 596, 627.

138. Ibid., pp. 658–659.

139. Ibid., pp. 636, 644.

140. Goddard, "Memorandum at the End of 1927 Diary," *Goddard Digital Collection.*

141. Goddard, *The Papers of Robert Goddard,* vol. 2, p. 648.

142. Ibid., p. 648. Buzz Aldrin contends that his father played an important role in connecting Goddard to Lindbergh. Although the records are unclear on the matter, given his father's extensive connections it is plausible and would be an extraordinary historical relationship if true. Keogh, J., "From Clark to the Moon," *CLARK Magazine,* Fall 2001.

143. Goddard, *The Papers of Robert Goddard,* vol. 2, p. 653.

144. Ibid., p. 660.

145. Ibid., p. 665.

146. Ibid., p. 674.

147. Ibid., p. 668.

148. Ibid., p. 685.

149. Ibid., p. 659.

150. Ibid., vol. 1, p. viii. For more on the Guggenheim support of American aviation, see Cleveland, R., *America Fledges Wings* (New York, 1942).

151. Clary, *Rocket Man*, p. 142.

152. Lindbergh, C., *Autobiography of Values* (Evanston, 1992), p. 15.

153. Goddard, *The Papers of Robert Goddard*, vol. 2, p. 715.

154. Ibid.

155. Ibid., p. 723.

156. Ibid., pp. 723–726.

157. Ibid., p. 725.

158. Unger, I., and Unger, D., *The Guggenheims: A Family History* (New York, 2005), pp. 85–86, 96, 114, 118–120.

159. Goddard, *The Papers of Robert Goddard*, vol. 2, p. 744.

160. Ibid., vol. 3, pp. 1556, 1670.

161. Ibid., vol. 2, p. 789.

162. Ibid., pp. 790, 820.

163. Ibid., p. 851.

164. Ibid., pp. 844–845.

165. Ibid., pp. 846–847.

166. Ibid., p. 864.

167. Ibid., pp. 830, 849, 874.

168. Ibid., p. 992.

169. Ibid.

170. For more on the contemporary development in Germany, see Neufeld, M. J., *The Rocket and the Reich: Peenemünde and the Coming of the Ballistic Missile* (New York, 1995), pp. 50–51.

171. Clary, *Rocket Man*, p. 174.

172. Lehman, *This High Man*, p. 239; Goddard, *The Papers of Robert Goddard*, vol. 2, pp. 1006, 1028–1029, 1034–1036, 1063, 1069, and vol. 3, pp. 1128–1129, 1133, 1134, 1177, 1199, 1205–1207.

173. Clary, *Rocket Man*, p. 162.

174. Ibid.

175. Goddard, *The Papers of Robert Goddard*, vol. 2, p. 852, and vol. 3, p. 1205.

176. Ibid., vol. 3, p. 1171.

177. Ibid., pp. 1132–1133, 1171, 1176–1177.

178. Ibid., p. 1127.

179. Clary, *Rocket Man*, p. 189.

180. Lehman, *This High Man*, p. 293.

181. Clary, *Rocket Man*, p. 200.

182. Goddard, *The Papers of Robert Goddard*, vol. 3, p. 1427.

183. Clary, *Rocket Man*, p. 173.

184. Goddard, *The Papers of Robert Goddard*, vol. 3, p. 1409.

185. Clary, *Rocket Man*, p. 200.

186. Goddard, "Diary, May 24, 1937," and "Diary, September 19, 1938," *Goddard Digital Collection;* Goddard, *The Papers of Robert Goddard*, vol. 3, pp. 1205, 1208, 1209.

187. Goddard, *The Papers of Robert Goddard*, vol. 3, p. 1208.

188. Goddard, "Diary, March 11, 1939," *Goddard Digital Collection;* Goddard, *The Papers of Rocket Goddard*, vol. 3, pp. 1231, 1237, 1239.

189. Goddard, *The Papers of Robert Goddard*, vol. 3, p. 136.

190. For histories of the NACA, see Ronald, A., *Model Research: The National Advisory Committee for Aeronautics, 1915–1958* (Washington, DC, 1985); Hansen, J., *Engineer in Charge: A History of the Langley Aeronautical Laboratory, 1917–1958* (Washington, DC, 1987).

191. Goddard, "Diary, May 24, 1937," *Goddard Digital Collection.*

192. Goddard, *The Papers of Robert Goddard*, vol. 3, p. 1079.

193. Ibid., pp. 1083, 1130, 1223.

194. Ibid., pp. 1108, 1216.

195. Ibid., p. 1396.

196. Ibid., pp. 1397–1403, 1406–1407.

197. Although Goddard's team built a functional liquid-fuel JATO, its utility paled in comparison to the solid-fuel JATO that the GALCIT group had developed. Ibid., pp. 1432–1433.

198. Ibid., p. 1502.

199. Rhodes, R., "The Ordeal of Robert Hutchings Goddard: 'God Pity a One-Dream Man,'" *American Heritage* 31, no. 4 (1980), pp. 25–32.

200. Clary, *Rocket Man*, p. 201.

201. Goddard later recouped these expenditures and repaid the $10,000 Guggenheim loan during the period of his contract. Goddard, *The Papers of Robert Goddard*, vol. 3, p. 1440.

202. Ibid., p. 1475.

203. It is worth noting here that although Goddard expended significant time and resources on his patents, they never played a significant economic role in his career. Indeed, he was wholly unaware of the storied history his most lucrative patent—his vacuum-tube radio oscillator—had enjoyed in lawsuit challenges to RCA until Arthur Collins approached him as part of his lawsuit with the radio giant. For these reasons, I support the proposition put forward by both Lehman and Clary that Goddard's extensive patenting was largely a result of his ego and concern for establishing priority.

204. Neufeld, M., *Von Braun: Dreamer of Space, Engineer of War* (New York, 2008), p. 54.

4. In the Eyes of the World

1. For a historiography of American spaceflight, see Siddiqi, A. A., "American Space History: Legacies, Questions, and Opportunities for Further Research," in *Critical Issues in Spaceflight,* ed. S. J. Dick and R. Launius (Washington, DC, 2006), pp. 433–480; Johnson, S., "The History and Historiography of National Security Space," in ibid., pp. 481–548.

2. Van Dyke, V., *Pride and Power: The Rationale of the Space Program* (Urbana, 1964).

3. McDougall, W. A., *The Heavens and the Earth: A Political History of the Space Age* (New York, 1985); Logsdon, J. M., *John F. Kennedy and the Race to the Moon* (New York, 2010).

4. President's Science Advisory Committee, *Introduction to Outer Space* (Washington, DC, 1958).

5. Emme, E., *The Impact of Air Power: National Security and World Politics* (Princeton, 1959), p. 844.

6. U.S. Congress, House Select Comm. on Astronautics and Space Exploration, *The United States and Outer Space,* H.R. Rep. No. 2710, 85th Cong., 2d sess. (1959), p. 6.

7. *Congressional Record* 107, no. 96 (June 8, 1961), p. 9174.

8. Office of the President, *Aeronautics and Space Report of the President: Fiscal Year 2012 Activities* (Washington, DC, 2012), http://history.nasa.gov/presrep2012 .pdf, p. 145.

9. Van Dyke, *Pride and Power,* p. 42.

10. Medaris J., *Countdown for Decision* (New York, 1960).

11. Bilstein, R., *Stages to Saturn: A Technological History of the Apollo/Saturn Launch Vehicles* (Washington, DC, 1996), p. 41.

12. U.S. Congress, Senate Comm. on Armed Services, *Military Procurement Authorization for Fiscal Year 1964,* Hearings, 88th Cong., 1st sess. (1963), p. 152.

13. U.S. Congress, House Comm. on Science and Astronautics, Subcommittee on Manned Space Flight, *1964 NASA Authorization,* pt. 2(a), Hearings, 88th Cong., 1st sess. (1963), p. 424.

14. Webb, J., and R. McNamara, "Recommendations for Our National Space Program: Changes, Policies, Goals," May 8, 1961, in *Exploring the Unknown: Selected Documents in the History of the U.S. Civil Space Program,* vol. 7 (Washington, DC, 2008), p. 492.

15. U.S. Congress, House Select Comm. on Astronautics and Space Exploration, *Astronautics and Space Exploration,* Hearings, 85th Cong., 2nd sess. (1958), p. 776.

16. For DuBridge's views five years later, see U.S. Congress, Senate Comm. on Aeronautical and Space Sciences, *Scientists' Testimony on Space Goals,* Hearings, 88th Cong., 1st sess. (1963).

17. Van Dyke, *Pride and Power,* p. 87.

18. *Congressional Record* 108, no. 24 (February 20, 1962), p. 2391.

19. Rechtin, E., "What's the Use of Racing for Space?" *Air Force/Space Digest* 44 (October 1961), p. 51.

20. Studies of the politics of science in America and elsewhere have been numerous. For especially well researched works, see Gilpin, R., and Wright, C., *Scientists and National Policy-Making* (New York, 1964); Dupree, A. H., *Science in the Federal Government* (Baltimore, 1985); Kealey, T., *The Economic Laws of Scientific Research* (London, 1997); Greenberg, D. S., *The Politics of Pure Science* (Chicago, 1999).

21. Divine, R., *The Sputnik Challenge: Eisenhower's Response to the Soviet Satellite* (Oxford, 1993).

22. Knorr, K., "On the International Implications of Outer Space," *World Politics* 12, no. 4 (July 1960), p. 578.

23. Alston, G., "International Prestige and the American Space Programme" (Ph.D. diss., University of Oxford, 1990).

24. Online Etymology Dictionary, "prestige," http://www.etymonline.com /index.php?search=PRESTIGE, accessed May 16, 2011.

25. Oliver, F. S., *The Endless Adventure,* vol. 2 (London, 1931), pp. 123–124.

26. Carr, E. H., *Great Britain as a Mediterranean Power* (Nottingham, 1937), p. 10.

27. Shimbori, M., Ikeda, H., Ishida, T., and Kondo, M., "An Attempt to Construct a National Prestige Index," *Indian Journal of Social Research* 3, no. 1 (1961), p. 25.

28. Morgentau, H., *Politics Among Nations: The Struggle for Power and Peace* (New York, 1948), pp. 77–78; Alston, "International Prestige and the American Space Programme," p. 17.

29. Berkner, L. V., "Space Research—A Permanent Peacetime Activity," in *Peacetime Uses of Outer Space,* ed. S. Ramo (New York, 1961), pp. 1–16.

30. Bartos, A., *Kosmos: A Portrait of the Russian Space Age* (Princeton, 2001), p. 92.

31. McDougall, W. A., *The Heavens and the Earth,* pp. 244–249.

32. Van Dyke, *Pride and Power,* p. 260.

33. Muir-Harmony, T., "Friendship 7's 'Fourth Orbit,'" 2012, http://blog.nasm .si.edu/space/friendship-7's-'fourth-orbit'/.

34. Offer, A., "Lecture Notes on Economic and Social History," Michaelmas term, University of Oxford, 2011.

35. Spence, M., "Informational Aspects of Market Structure: An Introduction," *Quarterly Journal of Economics* 90, no. 4 (November 1976), p. 592.

36. Douglas Aircraft Company, Inc., "Preliminary Design for an Experimental World-Circling Spaceship," RAND Corporation Rep. No. SM-11827, May 2 (Santa Monica, 1946), p. 2.

37. RAND, *RAND: 25th Anniversary Volume* (Santa Monica, 1973), p. 7.

38. Douglas Aircraft., "Preliminary Design for an Experimental World-Circling Spaceship," p. 219.

39. Van Dyke, *Pride and Power,* p. 13.

40. Grosse, A., "Report on the Present Status of the Satellite Problem," August 25, 1953, in *Exploring the Unknown: Selected Documents in the History of the U.S. Civil Space Program,* vol. 1 (Washington, DC, 1995), p. 268.

41. U.S. Congress, Senate Comm. on Armed Services, Preparedness Investigating Subcommittee, *The United States Guided Missile Program,* 86th Cong., 1st sess. (1959), p. 99; Van Dyke, *Pride and Power,* p. 11.

42. The Vanguard program was expected to cost $20 million–$30 million, a nontrivial $846 million–$1.27 billion in 2015 relative GDP-ratio adjusted equivalent value. It ended up costing $110 million. Portree, D., *NASA's Origins and the Dawn of the Space Age* (Washington, DC, 1998).

43. Bille, M., and Lishock, E., *The First Space Race: Launching the World's First Satellites* (College Station, 2004), p. 107.

44. Baker, D., *The Rocket: The History and Development of Rocket & Missile Technology* (New York, 1978), p. 133.

45. The psychology concepts of loss aversion and status-quo bias can be helpful in interpreting the decision-making dynamics of this period. See Kahneman, D., and Tversky, A., "Prospect Theory: An Analysis of Decision under Risk," *Econometrica* 47 (1990), pp. 263–291, and Samuelson, W., and Zeckhauser, R. J., "Status Quo Bias in Decision Making," *Journal of Risk and Uncertainty* 1 (1988), pp. 7–59.

46. Almond, G., "Public Opinion and the Development of Space Technology," *Public Opinion Quarterly* 24, no. 4 (1960), p. 571.

47. Brzezinski, M., *Red Moon Rising: Sputnik and the Hidden Rivalries That Ignited the Space Age* (New York, 2008), p. 180.

48. Eisenhower, D. D., "Are We Headed in the Wrong Direction?" *Saturday Evening Post,* August 11–18, 1962, pp. 19–25.

49. USIA, Office of Research and Intelligence, "World Opinion and the Soviet Satellite: A Preliminary Evaluation," in *NASA's Origins and the Dawn of the Space Age,* ed. D. Portree (Washington, DC, 1998), pp. 21–26.

50. Almond, G., "Public Opinion and the Development of Space Technology: 1957–1960" in *Outer Space in World Politics,* ed. J. Goldsen (New York, 1963), pp. 71–96.

51. Ibid., p. 76.

52. Ibid., p. 77.

53. Ibid., p. 88.

54. Lubell, S., "Sputnik and American Public Opinion," *Columbia University Forum* 1, no. 1 (Winter 1957), p. 15.

55. McLaughlin Green, C., and Lomask, M., *Vanguard: A History* (Washington, DC, 1970), p. 210.

56. Cadbury, D., *Space Race: The Epic Battle Between America and the Soviet Union for Dominion of Space* (New York, 2007), p. 173.

57. Eisenhower, D. D., "Official White House Transcript of President Eisenhower's Press and Radio Conference #123," concerning the development by the United States of an Earth satellite, October 9, 1957, https://www.eisenhower.archives.gov /research/online_documents/sputnik/10_9_57.pdf.

58. Alston, "International Prestige and the American Space Programme," p. 83.

59. U.S. Congress, House Comm. on Science and Astronautics, *Missile Development and Space Sciences,* Hearings, 86th Cong., 1st sess. (1959), p. 20.

60. U.S. Congress, House Committee on Science and Astronautics, *Review of the Space Program: Hearing Before the Committee on Science and Astronautics,* no. 3, pt. 1, 86th Cong., 2nd sess. (January 22, 1960), pp. 36–37.

61. Van Dyke, *Pride and Power,* p. 123.

62. Smith, D., *Communication via Satellite: A Vision in Retrospect* (Boston, 1976), p. 50. The full message: "This is the President of the United States speaking. Through the marvels of scientific advance, my voice is coming to you from a satellite circling in outer space. My message is a simple one. Through this unique means, I convey to you and all mankind America's wish for peace on earth and good will to men everywhere."

63. "Premier Calls His First Hot Dog a World Beater; PREMIER PRAISES HIS FIRST HOT DOG," *New York Times,* September 23, 1959, p. 1.

64. McDougall, *The Heavens and the Earth,* p. 202.

65. Van Nimmen, J., and Bruno, L., *NASA Historical Data Book: Volume I— NASA Resources 1958–1968* (Washington, DC, 1988), p. 129.

66. U.S. Congress, Senate, Report of the Committee on Commerce, *The Speeches, Remarks, Press Conferences, and Statements of Senator John F. Kennedy, August 1 Through November 7, 1960,* S. Rep. No. 994, pt. 1, 87th Cong., 1st sess. (1961), pp. 159, 377.

67. Edward Smith, J., "Kennedy and Defense: The Formative Years," *Air University Review,* March–April 1967.

68. Kennedy, J. F., "Remarks at a Meeting with the Headquarters Staff of the Peace Corps," June 14, 1962, in *The American Presidency Project,* ed. J. T. Woolley and G. Peters (online), available at http://www.presidency.ucsb.edu/ws/index.php ?pid=8718.

69. Kissinger, H., *Nuclear Weapons and Foreign Policy* (New York, 1957).

70. Kissinger, H., "Military Policy and the Defense of the 'Grey Areas,'" *Foreign Affairs* 33, no. 3 (1955), pp. 416–428.

71. Kraus, S., "Televised Presidential Debates and Public Policy" (New York, 2000), debate transcripts available at http://www.presidency.ucsb.edu/debates.php.

72. Alston, "International Prestige and the American Space Programme," p. 212.

73. Nixon, R., "Remarks of the Vice President of the United States, Park Forest, IL," October 29, 1960, in *The American Presidency Project,* available at http://www.presidency.ucsb.edu/ws/?pid=25513.

74. "Nixon Calls Missile Lag 'Inherited'" *St. Petersburg Times,* October 26, 1960, p. A-9.

75. Finney, J., "U.S. Claims Scientific Lead Despite the Soviet Union's 'Spectaculars,'" *New York Times,* April 16, 1961, p. E3.

76. Logsdon, J., *The Decision to Go to the Moon: Project Apollo and the National Interest* (Cambridge, 1970), pp. 111–112.

77. Johnson, L. B., "Evaluation of Space Program," April 28, 1961, *Exploring the Unknown: Selected Documents in the History of the U.S. Civil Space Program,* vol. 1 (Washington, DC, 1995), pp. 427–429.

78. Ibid., p. 427.

79. Webb and McNamara, "Recommendations for Our National Space Program," p. 492.

80. Kennedy, J. F., "Address in Los Angeles at a Dinner of the Democratic Party of California," December 18, 1961, in *The American Presidency Project,* available at http://www.presidency.ucsb.edu/ws/index.php?pid=8452.

81. Kennedy, J. F., "Address at Rice University on the Nation's Space Effort," September 12, 1962, Houston, TX, available at http://explore.rice.edu/explore/Kennedy_Address.asp.

82. Kennedy, J. F., "Special Message to the Congress on Urgent National Needs," May 25, 1961, Public Papers of the Presidents, John F. Kennedy, 1961, doc. 205, pp. 396–405, available at www.nasa.gov/pdf/59595main_jfk.speech.pdf.

83. Ibid.

84. Ibid.

85. U.S. Congress, House Comm. on Science and Astronautics, *Review of the Space Program,* pt. 1, 86th Congress., 2d sess. (1960), p. 30.

86. McNamara, R. "Brief Analysis of Department of Defense Space Program Efforts," April 21, 1961, *Exploring the Unknown: Selected Documents in the History of the U.S. Civil Space Program,* vol. 1 (Washington, DC, 1995), p. 424.

87. Harvey, B., *Soviet and Russian Lunar Exploration* (New York, 2007), pp. 52–54.

88. U.S. Congress, *Miscellaneous Reports on Public Bills II,* vol. 2, H.R. Rep., 85th Cong., 2nd sess. (1958), p. 19.

89. Ibid., p. 22.

90. Johnson, L. B., "Special Message to the Senate on Transmitting the Treaty on Outer Space," February 7, 1967, http://www.presidency.ucsb.edu/ws/?pid=28427.

91. U.S. Congress, House Comm. on Science and Astronautics, *The Practical Values of Space Exploration* (rev. August 1961), H. Rep. No. 1276, 87th Cong., 1st sess. (1961), p. 22.

92. Ezell, E., and Ezell, L., *The Partnership: A History of the Apollo-Soyuz Test Project* (Washington, DC, 1978), pp. 39–44.

93. Finley, D., "Soviet-U.S. Cooperation in Space and Medicine: An Analysis of the Détente Experience," in *Sectors of Mutual Benefit in U.S.-Soviet Relations,* ed. N. Jamgotch (Durham, 1985), pp. 133–136.

94. Karash, Y., *The Superpower Odyssey: A Russian Perspective on Space Coopera-tion* (Reston, 1999), pp.105–106. The base year for calculation purposes is 1975.

95. Ibid., p. 114.

96. Frutkin, A. W., "International Programs of NASA," in *The Challenges of Space,* ed. H. Odishaw (Chicago, 1962), p. 273.

97. Veblen, T., *The Theory of the Leisure Class: An Economic Study of Institutions* (New York, 1899), p. 75.

98. Ibid., p. 102.

99. Ibid., p. 99.

100. Launius, R., "Public Opinion Polls and Perceptions of US Human Space-flight," *Space Policy* 19 (2003), pp. 163–175.

101. For the full story, see Heppenheimer, T. A., *The Space Shuttle Decision 1965–1972* (Washington, DC, 2002).

102. Fletcher, J. C., "The Space Shuttle," November 22, 1971, in *Exploring the Unknown: Selected Documents in the History of the U.S. Civil Space Program,* vol. 1 (Washington, DC, 1995), pp. 555–556.

103. Weinberger, C., "Memorandum for the President: Future of NASA," August 12, 1971, in ibid., pp. 546–547.

104. Ibid., p. 547.

105. Heppenheimer, *The Space Shuttle Decision 1965–1972,* p. 392.

106. Low, G., Deputy Administrator, NASA, Memorandum for the Record, "Meeting the President on January 5, 1972" January 12, 1972, NASA Historical Reference Collection, History Office, NASA Headquarters, Washington, DC.

107. McCurdy, H., *The Space Station Decision: Incremental Politics and Techno-logical Change* (Baltimore, 1990); Logsdon, J., *Together in Orbit: The Origins of International Participation in the Space Station* (Washington, DC, 1998).

108. Reagan, R., "Address Before a Joint Session of the Congress on the State of the Union," January 25, 1984, in *The American Presidency Project,* available at http://www.presidency.ucsb.edu/ws/index.php?pid=40205.

109. Church, G., "Reagan Gets Ready," *Time Magazine,* January 30, 1984.

110. Reagan, R., and Bush, G., 1984 Campaign Brochure, "Leadership That's Work-ing," http://www.4president.org/brochures/1984/reaganbush1984brochure.htm.

111. Ibid.

112. Ibid.

113. Reagan, R., "Remarks at a Reagan-Bush Rally in Fairfield, Connecticut," Oc-tober 26, 1984, available at http://www.reagan.utexas.edu/archives/speeches/1984/102684b.htm.

114. Ibid.

115. Smith, M., "NASA's Space Station Program: Evolution and Current Status—Testimony Before the House Science Committee, April 4, 2001," Congressional

Research Service (Washington, DC, 2001), http://history.nasa.gov/isstestimony2001
.pdf.

116. Clinton. W. J., "Address Before a Joint Session of Congress on Administration Goals," February 17, 1993, in *The American Presidency Project,* available at http://www.presidency.ucsb.edu/ws/index.php?pid=47232.

117. Ibid.

118. Clinton. W. J., "Statement on the Space Station Program," June 17, 1993, in *The American Presidency Project,* available at http://www.presidency.ucsb.edu/ws /index.php?pid=46709.

119. Ibid.

120. Ibid.

121. USIA, Office of Research and Intelligence, "World Opinion and the Soviet Satellite," p. 23; Almond, G., "Public Opinion and the Development of Space Technology: 1957–1960," pp. 71–96. For a full exploration of the ambiguities, see Bainbridge, W. S., "The Impact of Space Exploration on Public Opinions, Attitudes, and Beliefs," *Historical Studies in the Societal Impact of Spaceflight* (Washington, DC, 2015), pp. 1–76.

122. Worden, S. P., *SDI and the Alternatives* (Washington, DC, 1991); Schweizer, P., *Victory: The Reagan Administration's Secret Strategy That Hastened the Collapse of the Soviet Union* (New York, 1994); Hertzberg, H., "Laser Show," *New Yorker,* May 15, 2000; Fitzgerald, F., *Way Out There in the Blue: Reagan, Star Wars, and the End of the Cold War* (New York, 2000); Hey, N., *The Star Wars Enigma: Behind the Scenes of the Cold War Race for Missile Defense* (Washington, DC, 2006).

123. Westwick, P., "From the Club of Rome to Star Wars: The Era of Limits, Space Colonization, and the Origins of SDI," (working paper, Envisioning Limits Conference: Berlin, April 21, 2012), p. 1.

124. Ibid., pp. 1–5.

125. McCurdy, H., *Inside NASA: High Technology and Organizational Change in the U.S. Space Program* (Baltimore, 1993).

126. Siddiqi, A. A., *Challenge to Apollo: The Soviet Union and the Space Race, 1945–1974* (Washington, DC, 2000).

127. Bainbridge, W., *The Spaceflight Revolution: A Sociological Study* (New York, 1976), p. 81.

128. DeGroot, G., *The Dark Side of the Moon: The Magnificent Madness of the American Lunar Quest* (New York, 2006).

129. Bainbridge, *The Spaceflight Revolution,* p. 1.

The Next Space Patrons

1. Clarke, A. C., *The Challenge of the Spaceship* (New York, 1960), p. 68.

2. Bromberg, J. L., *NASA and the Space Industry* (Baltimore, 2000).

3. "Constitution of the Cincinnati Astronomical Society," in Adams, J. Q., *An Oration Delivered Before the Cincinnati Astronomical Society* (Cincinnati, OH, 1843), p. 70.

INDEX